In Praise of Imperfection

BOOKS IN THE ALFRED P. SLOAN FOUNDATION SERIES

Disturbing the Universe *by Freeman Dyson*
Advice to a Young Scientist *by Peter Medawar*
The Youngest Science *by Lewis Thomas*
Haphazard Reality *by Hendrik B. G. Casimir*
In Search of Mind *by Jerome Bruner*
A Slot Machine, a Broken Test Tube *by S. E. Luria*
Enigmas of Chance *by Mark Kac*
Rabi: Scientist and Citizen *by John Rigden*
Alvarez: Adventures of a Physicist *by Luis W. Alvarez*
Making Weapons, Talking Peace *by Herbert F. York*
The Statue Within *by François Jacob*

THIS BOOK IS PUBLISHED AS PART OF AN ALFRED P. SLOAN FOUNDATION PROGRAM

IN PRAISE
OF
IMPERFECTION

My Life and Work

RITA LEVI-MONTALCINI

Translated by Luigi Attardi

Basic Books, Inc., Publishers   New York

Library of Congress Cataloging-in-Publication Data

Levi-Montalcini, Rita.
   In praise of imperfection.

   Translation of: Elogio dell'imperfezione.
   1. Levi-Montalcini, Rita.   2. Neurologists
—Italy—Biography.   3. Neurologists—United
States—Biography.   I. Title.
RC339.52.L48A3   1988   616.8'092'4   [B] 87–47781
ISBN   0–465–03217–6

To Paola

and to the memory of our father

whom she adored while he lived

and whom I loved and worshiped

after his death

The Choice

The intellect of man is forced to choose

Perfection of the life, or of the work,

And if it take the second must refuse

A heavenly mansion, raging in the dark.

When all that story's finished, what's the news?

In luck or out the toil has left its mark:

That old perplexity an empty purse,

Or the day's vanity, the night's remorse.

—William Butler Yeats

CONTENTS

Preface to the Series   xi
Acknowledgments   xiii

Introduction   3

PART I
HEREDITY AND ENVIRONMENT

1 Turin: Royal City and Home Town   11
2 Freethinker, with Misgivings   18
3 Two X Chromosomes in a Victorian Climate   26
4 The Choice   33
5 The Death of My Father   42
6 Medical Student: A Master's Apprentice   48
7 Fellow Students: Friends for Life   62

PART II
THE DIFFICULT YEARS

8   Premonitions of Trouble   73
9   The Start of the Anti-Semitic Campaign: The
    Racial Manifesto   79
10 A Private Laboratory *à la* Robinson Crusoe   88
11 Life in Hiding   99
12 Caring for War Refugees   106
13 The Return to Turin and to Research   110

CONTENTS

PART III
A NEW LIFE

14 A New Continent   117
15 Experimental Neurobiology in the First Half of
   the Century   123
16 Explorations: The Midwest and the Developing
   Nervous System   135
17 The Nerve Growth Factor: The Opening of the
   Saga   144
18 The Fibrillar Halo and Carnival in Rio   153
19 Stan and the NGF   161
20 In Memory of a Friend   169

PART IV
BACK TO MY NATIVE COUNTRY

21 The Tug of Early Affections   185
22 The Miracle of Maxwell's Demon   192
23 Farewell to a Master—and to a Father   202
24 Disharmony in a Complex System   206
25 Epilogue: Primo Levi's Message   212

Notes   215
Index   217

Photographs   *following* 112

PREFACE TO THE SERIES

THE ALFRED P. SLOAN FOUNDATION has for many years had an interest in encouraging public understanding of science. Science in this century has become a complex endeavor. Scientific statements may reflect many centuries of experimentation and theory, and are likely to be expressed in the language of advanced mathematics or in highly technical terms. As scientific knowledge expands, the goal of general public understanding of science becomes increasingly difficult to reach.

Yet an understanding of the scientific enterprise, as distinct from data, concepts, and theories, is certainly within the grasp of us all. It is an enterprise conducted by men and women who are stimulated by hopes and purposes that are universal, rewarded by occasional successes, and distressed by setbacks. Science is an enterprise with its own rules and customs, but an understanding of that enterprise is accessible, for it is quintessentially human. And an understanding of the enterprise inevitably brings with it insights into the nature of its products.

The Sloan Foundation expresses great appreciation to the advisory committee. Present members include the chairman, Simon Michael Bessie, Co-Publisher, Cornelia and Michael Bessie Books; Howard Hiatt, Professor, School of Medicine, Harvard University; Eric R. Kandel, University Professor, Columbia University College of Physicians and Surgeons and Senior Investigator, Howard Hughes Medical Institute; Daniel Kevles, Professor of History, California Institute of Technology; Robert Merton, University Professor Emeritus, Columbia University; Paul Samuelson, Institute Professor of Economics, Massachusetts Institute of Technology; Robert Sinsheimer, Professor of Biophysics, California Institute of Technology; and Steven Weinberg, Professor of Physics, University of Texas at Austin; Stephen White, former Vice-President of the Alfred P. Sloan Foundation. Previous members of the committee were Daniel McFadden, Professor of Economics, and Professor Philip Morrison, Professor of Physics, both of the Massachusetts Institute of Technology; Mark Kac (deceased), formerly Professor of Mathematics, University of Southern California; and Frederick E. Terman [deceased], formerly Provost Emeritus, Stanford University. The Sloan Foundation has been represented by Arthur L. Singer, Jr., Stephen White, Eric Wanner, and Sandra Panem. The first publisher of the program, Harper & Row, was represented by Edward L. Burlingame and Sallie Coolidge. This volume is the fifth to be published by Basic Books, represented by Martin Kessler and Richard Liebmann-Smith.

—ALBERT REES
President
Alfred P. Sloan Foundation
March 1988

ACKNOWLEDGMENTS

THE AUTHOR wishes to express her gratitude to the Alfred P. Sloan Foundation for the prompting and assistance it provided, and to her dear friend Luigi Attardi for his excellent translation, valued suggestions, and cherished association in the drafting of this life history.

In Praise of Imperfection

Introduction

A YOUNG TAIWANESE assistant of mine, who had come to Italy to enroll in the University of Perugia's department of biology about ten years before we met, quickly learned to speak the language and followed his degree courses without great difficulty. He expressed himself well, though not only his pronunciation but his choice of adjectives and syntax continued to reveal his foreign origin. One day, while observing two nests of mice which differed from each other in the respective mother's behavior, in health, and number of offspring, he smiled approvingly at the mouse suckling the little ones pressing around her, and called her "Mamma Carissima" (dearest mother). Then, scowling severely at the other who was entirely neglecting her young—few in number because, as often happens among rodents, she had eaten many of them after giving birth—he called her "Mamma Sporchissima" (filthy mother). No Italian would ever have used these terms even when sharing my assistant's preference for the nursing mother. It is hard to say whether his use of the adjectives *dearest* and *filthy* was due to insufficient knowledge of the language or to an ethical—rather than, as in our case, merely biological—understanding of the behavior of mice and other animals. Perhaps differing from Orientals in this respect, Westerners believe that our species enjoys the privilege of behaving according to ethical principles, even while we acknowledge that only a few people actually apply them. Johnny, as we used to call our Taiwanese collaborator, adopting the affectionate nickname used by his young Sienese wife and fellow student, would also amuse us when he declared, "You Italians, all the same. We Chinese, all different." We, in comparing Italians and Chinese, apply his same criteria but arrive at opposite conclusions. Both claims derive from the optic system's characteristic of picking out at first glance the most noticeable features of the animate or inanimate objects that enter one's field of vision. In our species, the immediate somatic differences are the color of skin and hair, whether the latter

3

is smooth or crinkly, the shape of the eyes, the height and general shape of the body. This set of visual perceptions enables us to determine immediately at first encounter whether an individual belongs to the white, the yellow, the black, or the so-called red-skinned race, the latter's skin being somewhat brick-colored. These somatic features—much more noticeable than others such as eye color and expression and facial features, which vary among individuals of the same race—justify Johnny's claim that *we* are all the same and *they* all different.

The human species not only has to rely mainly upon the optic system, which played and continues to play an essential role in the preservation of both individuals and species, but is also highly emotive. Emotions experienced during the first years of life and sensations that have caused one joy or pain leave indelible traces which condition a person's actions and reactions throughout life. Moreover, we blame on these emotions and sensations aberrations in a person's conduct whenever it deviates from the rules of good behavior. Even if this explanation of conduct frees individuals from all sense of guilt—which falls automatically on parents or their surrogates—it does not necessarily free one from the guilt complexes that remain despite the absolution offered by Freud. I consider it my great, if not my greatest, good fortune not to have suffered from such complexes. The reason is, I think, that I was brought up in an environment that, though not permissive, was brimming with affection and never troubled by disagreements between my mother and father. The possible negative influence of having been born and raised in a Victorian climate, unsuited to my natural tendencies, was mitigated by my mother's complete acceptance of the role prescribed for women during the reign of Queen Victoria and the first two decades of this century. The absence of complexes, a remarkable tenacity in following the path I believe to be right, and a way of underestimating the obstacles standing between me and what I want to accomplish—a trait I believe I inherited from my father—have helped me enormously in facing the difficult years of life.

To him, as to my mother, I owe a tendency to look on others with sympathy and without animosity and to see things and people in a favorable light. This attitude, common to all of my parents' children, showed itself most powerfully in my brother Gino, making a strong impression on me as a child and determining at least in part my unconditional admiration for him. More even than to me, this attitude was of great benefit to him throughout a life that was ended abruptly by a heart attack while he was still fully active. Contemplating the sunset from the pine grove in Forte dei Marmi where my sisters and I spent the month of August with him and his family, he would say, "Life is all a miracle." That sense of amazement before the immensely complex phenomenon of life absorbed him entirely and was the reason his serenity was never troubled by resentment for injustices suffered or a lack of gratitude in those he helped.

Looking back now on the long path my life has followed, on the lives of my peers and colleagues, and on the briefer ones of the young recruits who have worked with us, I have become persuaded that, in scientific research, neither the degree of one's intelligence nor the ability to carry out one's tasks with thoroughness and precision are factors essential to personal success and fulfillment. More important for the attaining of both ends are total dedication and a tendency to underestimate difficulties, which cause one to tackle problems that other, more critical and acute persons instead opt to avoid.

Without pre-established plan and guided at every turn rather by my inclinations and by chance, I have tried—as will be clear from a reading of this sort of balance-sheet or final account of my life—to reconcile two aspirations that the Irish poet William Butler Yeats deemed to be irreconcilable: perfection of the life and perfection of the work. By so doing, and in accordance with his predictions, I have achieved what might be termed "imperfection of the life and of the work." The fact that the activities that I have carried out in such imperfect ways have been and still are for me a source of inexhaustible joy, leads me to believe that imperfection, rather than perfection, in the execution of our assigned or elected tasks is more in keeping with human nature. The evolutionary processes that acted on the descendants of our ancestress Lucy, who lived three and a half million years ago, and transformed that forty-one-inch-tall female humanoid with a skull no larger than a cocoanut into *Homo sapiens,* were not preordained but were, rather, the fruit of chance mutations. The progressive increase in the volume of the brain and the even more spectacular increase in its intellectual capacities are the result of an unharmonious process which has generated in us an infinite amount of psychic complexes and behavioral aberrations. These aberrations have been spared our traveling companions, the anthropomorphic primates, or those infinitely more numerous and humble creatures who preceded us by millions of years and will yet outlive us: the insects. The insects inhabiting the surface of our planet today are not substantially different from their remote ancestors who flourished six hundred million years ago. Since the first specimen appeared, their speck of a brain has shown itself so well suited to the problems of the environment and of predators that it has not had to submit to the capricious game of mutations, but owes its evolutionary immobility to the perfection of a primordial model.

Similarly, the creativity of the brain of *Homo sapiens* expresses itself by elaborating either simple and perfect mechanical devices that so fulfill their purpose as not to require modification, or more rudimentary and imperfect ones that, by virtue of their very imperfection, lend themselves to being restructured. A simple and perfectly efficient means of locomotion, the bicycle, which is made up of two wheels connected by a transmission system and activated by the motor energy of the user's muscles, has shown itself, since the time of its invention in the middle of the last century, to be so entirely adequate to performing the work it was conceived for that only slight modifications have been brought to the primordial model. The history of another mechanical device, one invented more than two centuries ago, in 1769, by the Frenchman Nicholas Cuénot and tested a few decades later, has instead been very different. The latter, a rudimentary three-wheeled device activated by a motor, represented the ancestral form of those vehicles known as automobiles. This primitive propulsion mechanism was substantially modified in the decades following its invention in such ways as to render it always faster and more safe. At the same time, it served as a model for the invention of another, even faster means of transport: the airplane. The evolution of the latter, since the day in 1908 when de Lagrange's prototype soared above the ground to the marvel and exultation of onlookers and travelled several dozen miles at a low altitude, became, after the combustion engine was replaced by the jet turbine, far more rapid and imposing than that of the automobile. Finally, the introduction of an entirely new principle, that of the rocket, revolutionized aircraft propulsion systems and enabled man to land on the moon.

The analogy between the evolution of the brain of vertebrates and that of propulsion mechanisms is based on a single yet fundamental trait that living forms and mechanical artifacts have in common.

Imperfect machines such as Cuénot's lent themselves to the game of selection just as the brain of the first vertebrate did when it appeared on our planet in the Carboniferous period between 300 and 400 million years ago. The empty vesicles of that ancestral brain, unlike the solid and compact mass of the brain of invertebrates, succumbed to the selective pressure that evolution has exerted on cerebral vesicles, giving birth to the infinite number of variations (mutations) in the brains of both living and extinct forms of vertebrate life. The most recent among these is the marvelous though far from perfect brain of *Homo sapiens*. The solid, compact, and perfect brain of invertebrates, on the other hand, has not bent to such selective pressure, a fact especially true of the model that may well be considered the most successful of all: that of the insects, who have propagated and come to dominate the depths and surface of the planet. Neither a Hitler nor an Einstein will ever be descended from them (for better or for worse) though their progeny will perhaps still exist hundreds of millions of years from now.

PART I

Heredity and Environment

CHAPTER 1

Turin: Royal City and Home Town

> Those autumns in which the long shadows, the tranquillity of the sky, all that mood of happiness and convalescence released by nature after the criminal violence of the spring and the debilitating fevers of summer, bring the occult beauty of Turin to its highest degree of expression. Then all the throng of marble and bronze statues, the great men who all year long stand motionless upon their low plinths amid the continuous coming and going of vehicles and pedestrians, step down painfully from their pedestals and, having stretched their limbs, set off cautiously toward the famous Piazza Castello where their mysterious conventicles take place. They gather there to sing in chorus, under the perfectly pure autumnal sky, the ineffable hymn to eternal fidelity and eternal friendship.
>
> —GIORGIO DE CHIRICO,
> *Paola Levi-Montalcini*

THUS, the great twentieth-century Italian painter Giorgio De Chirico described his, and my, native city of Turin. He continued with his description:

> There one can see Garibaldi himself, the fearless soldier, the bearded lion with the eyes of a sentimental young girl, listening to Giuseppe Verdi telling him, in a low voice cracked with emotion, how he composed the famous aria sung by the baritone in the second act of *Il Trovatore*. ". . . il balen del suo sorriso. . . ." One can see King Vittorio Emanuele II, all in bronze, covered in ribbons, sashes and decorations, discussing strategy with Emanuele Filiberto of Savoia, leaning upon the hilt of his long sword; and all around, throughout the whole city, there is silence, happiness, and meditation. From the Wallace fountains, in the public squares, flows water cool and clear. On the façades of the stations, the hands of

11

clocks read two in the afternoon. The locomotives are resting, and above the roofs of the public buildings and of the great bazaars, oriflammes in soft and blazing colors flutter gently in the cool gusts which come from far off, from the depth of the valley, from the Alps which can be seen far away on the line of the clear horizon, their peaks forever capped with snow. Turin lives under the sign of the bull. The first inhabitants had the bull for emblem. They were the Taurins *(Taurinorum gens),* from which comes Turin. Now everybody knows that the bull is one of the four most enigmatic animals of creation. The others are the ass, the rooster, and the hen. All four of these animals are profoundly anthropomorphic. It is not for nothing that in Greek mythology a man with a bull's head is set to guard the most mysterious construction in the world of legend.

To my infant eyes, the monarchic and fluvial city of Turin did not appear quite as fascinating as it did to De Chirico. During the long winters, I would look out onto a large avenue from the window of our fourth-floor residence, and contemplate the barren plane trees laden with snow and beyond them, in a nearby square, the monument to Vittorio Emanuele II, first king of Italy by the grace of God and the people's wish, showing him "all in bronze, covered in ribbons, sashes and decorations." His gigantic figure which, unlike those of the other national heroes, rested on a pedestal taller than the surrounding nineteenth-century buildings, stood out against the gray winter sky with the majesty proper to the king who, in 1870, brought about national unity. His satisfaction with the task he had accomplished was conveyed by his haughty, frowning gaze, and even more by the immense mustache decorating his upper lip. Its prominence, evident even in daguerreotypes of the period, had been purposefully emphasized by the sculptor as a symbol of the king's virility. And yet it was not necessary to accentuate that mustache almost to the point of caricature to remind his subjects of their first king's masculine attributes. These were all too well known, at least to the Turinese at the end of the last century, who took pleasure not so much in his warlike exploits as in his famous hunting parties, when he indulged in great drinking bouts and in liberties with the beautiful local girls. One of them, *la bella Rosina,* had become legendary in Turin and was still talked about during my childhood when people recalled how she had bewitched the king and deposed the real queen through a morganatic marriage.

This picturesque aspect of the life of Vittorio Emanuele II, first king of Italy, contrasted with the traditional severity of the House of

Turin: Royal City and Home Town

Savoy and the ascetic tendencies of his immediate predecessor, Carlo Alberto, king of Piedmont and Sardinia. Vittorio Emanuele's well-known escapades had given rise to the rumor that he was the changeling of a scullery maid—a rumor reflecting the belief among well-to-do Turinese at the turn of the century that one is born son of scullery maid or king, one does not become either. In any case, Vittorio Emanuele II had certainly become king after his death, as was demonstrated by the haughty expression, the mustache, the puffed-out chest, and all those sashes, medals, and decorations. That famous mustache, which I contemplated from my window, was of such dimensions, my mother would say, that each curve could have served as an armchair for a grown man. Those curves impressed me also because they were similar to my father's, even though his were described as *à la Umberto:* that is, they were of the same cut as the mustache of the son of Vittorio Emanuele who succeeded him and went down in history, for lack of more salient characteristics, as "the good king." The mustaches of father and son had marked the age, no less than had crinolines and the wasp waists of the ladies. Both emphasized secondary sexual characteristics with the express purpose of underlining differences in role.

When I was still only a very small child, I developed an aversion to mustaches, and justified my reluctance to kiss my father by adducing the reason, in large part true, that they pricked me. "Rita," my father would note with poorly concealed disappointment, "doesn't know how to kiss. She would rather kiss the air than her father." I had, in fact, developed the habit, when approaching him for the good-night kiss, of turning my head away upon contact with his face and sending the kiss into the air. A person with such a sharp eye as my father certainly could not have missed that I had no difficulty at all in kissing Mother; but in her case not only the fondness I felt but also the pleasant contact with her soft and fragrant face were reason enough for me to overcome my natural reluctance to engage in physical contact. The question of aerial kisses, however, had also presented itself in her regard, though in a different way. In an essay in which our second-grade teacher had asked us to explain what fingers were for, I wrote—to the great amusement of my brother Gino and those who knew the story—that they were "for sending kisses to Mother." Those, too, were a form of aerial kiss but had, in her case, a well-defined target. They revealed my lack of practical sense which was to persist in the future. This habit of blowing kisses, which lasted—not only in my father's regard, and even in the absence of mustaches—into adolescence, ought to have revealed to a person of his sensitivity and intuition much about the personality of this daughter with whom he could not manage to establish the close relationship he had with my two sisters.

My twin, Paola, who adored him, had shown from early infancy a great artistic talent which aroused in me unconditional admiration, unsullied by envy or regret, because I was completely lacking in her gift. This was only one of the differences between us evident from the first years of our lives. The others, no less significant and which revealed at first glance our non-identical twinship, were manifest in our physical appearance, in our characters, and in our behavior. Her face differed from my own in shape and in its every feature. Beneath a high, slightly convex forehead, her laughing blue eyes denoted a disposition (in truth, more apparent than real) to a gaiety that enchanted our father. From an early age, which prevented one from guessing the still hidden design of her genes in the modeling of her facial features (a design puberty revealed to be entirely in keeping with one's expectations), her face bore an extraordinary resemblance to his own—a cause for joy and paternal pride. On the other hand, our mother was pleased to assert that I was the living image of her mother. She saw again in mine her mother's gray-green eyes with their melancholy gaze, the slight asymmetry of the face, the slender lines of the bone structure. My tendency to seek solitude and to flee from encounters with either sex reminded her of the sad, reserved character of her own mother whom she had adored and lost in adolescence. The deep affection linking Paola to our father, reciprocated with the most heartfelt tenderness, and that between myself and my mother date back to that early time. As for the relationship between Paola and myself, from our earliest childhood up to today—a period spanning over three quarters of a century—it has been characterized by an intensity of affection so great as to have created, especially while we were children, a sort of barrier against the intrusion of third parties.

Gino, our brother, seven years older than I, and Anna, five years older, came into the category of third parties. The barrier that excluded them from our secrets was to fall during adolescence when age differences were neutralized by the cultural inclinations and interests Paola shared with Gino and I with my elder sister. Gino had in common with Paola a strong artistic personality which made him choose architecture as a career. His natural talents manifested themselves in an exceptional aptitude for drawing—already evident, as in Paola, from childhood—and in a passion for modeling clay and plasticine (the smell of this malleable, gray-green, rubbery substance, which I believe can no longer be found in the shops, was noticeable on everything he touched and left permanent traces—among the most pleasant—on my olfactory neurons); these talents led him, after he had completed his high school studies with honors, to want to become a sculptor. Our father, however, whose dream it was that Gino would follow in his footsteps and take a degree in engineering, opposed a choice that offered no guarantees for the future. He yielded somewhat grudgingly to his son's clear preference for architecture, which offered a compromise between sculpture and engineering, the latter having no attraction for Gino whatsoever. The choice, as Gino himself later acknowledged, was a happy one. His natural artistic talents combined with passionate dedication made him one of the most prominent Italian architects of the mid-century and, at the same time, allowed him to express in his architectural constructions the innate sensibility of the sculptor. Our sister Anna (or Nina, as she was known to us and to friends) had literary interests similar to my own; yet neither she nor I were ever to fulfill our vague adolescent aspirations. In her case, marriage, pregnancies following one after another, and maternal duties took her away from the career she dreamed of: to become a writer like the Swedish Selma Lagerlöf, her favorite author and chosen model. Anna had passed on to me an unbounded admiration for Lagerlöf's books, and together we dreamed of the long nordic winters and shivered at the reading of Swedish sagas. Lagerlöf's 1891 novel *Gösta Berling* was our favorite subject of conversation, until we discovered in Emily Brontë's *Wuthering Heights* and Virginia Woolf's *To the Lighthouse* other models to emulate.

The alliance established between Paola and myself during our early childhood mitigated but did not free me from the anxieties I suffered as a child. These had their roots in extreme timidity, lack of self-confidence, and fear of adults in general and of my father in particular, as well as of monsters that might suddenly pop out of the dark and throw themselves upon me. To reach our bedroom and bathroom in the apartment where we then lived, one had to walk the whole length of a long hallway connecting the playroom of our pre-school days to the bedrooms. The moment dusk began to fall, ignoring Gino's joking remarks, I would ask Paola, who didn't suffer from these torments, to accompany me whenever I had to face the ordeal. This fear of the dark, and of malevolent beings who might take advantage of it to attack me, was not the only manifestation of my insecurity and anxious nature. A form of terror of which I have a vivid memory was that caused by the motion of wind-up toys. One of these arrived as a present from my father's sister, who lived in London and enjoyed great prestige in the eyes of us children because she spoke English and had married—as Mother put it—a high-class journalist. This particular toy was a small English gentleman wearing a black bowler hat, such as is still in fashion many decades later and was immortalized in photographs of Chamberlain, Eden, and Churchill. Our gentleman, no more than twenty-five centimeters tall, had a handsome mustache and a starched collar and held, in his gloved right hand, a walkingstick which he shook whenever he was set in motion by winding a little handle. The Britishness of his appearance was emphasized by a flannel jacket in black and white check (very tweedy, Agatha Christie would have said) and by the slow and dignified manner in which he advanced. When Gino wound him up and maliciously sent him toward me, panic would overwhelm me, and I would back away to re-establish distance between myself and "him"—to the great amusement of Gino and the family. "And yet," maintained my mother, who always protected me, "Rita is not easily frightened." And she would recall my easy acquaintance with animals much larger than myself and the pleasure I took in provoking a little goat in the country, exulting whenever it butted me and knocked me down. I think that the difference in my behavior derived from the fact that the little goat was a living creature. Fortunately for me, neither electronic mechanisms nor the possibility of activating machines or puppets by remote control had yet been thought of. I don't know how I would have reacted in the face of one of these if some force entirely mysterious to me had sent one in my direction.

Even if from early childhood I was bound by a far more vivid affection to my mother than to my father, it was he rather than she who had a decisive influence on the course of my life, both by transmitting to me a part of his genes and eliciting my admiration for his tenacity, energy, and ingenuity; and, at the same time, by provoking my silent disapproval of other aspects of his personality and behavior (a subject to which I shall return in chapter 3). From him, too, I inherited seriousness and dedication to work, and a secular, Spinozan conception of life. The difficulty my father and I had in communicating—a cause of pain to him no less than to me—continued until death cut him off prematurely when Paola and I had just turned twenty-three. His sudden death left an ineradicable mark on Paola who had adored him, and filled me with a sense of regret and remorse for having so disappointed him in being what he would tenderly define, trying to overcome the barrier between us, as his "shrinking violet." Given our fundamental similarities, I think that today he would recognize me as his daughter and not only as his "shrinking violet," and that the harmony missing in the difficult years of my childhood and adolescence would immediately spring up between us, repaying us for the suffering we involuntarily caused each other.

CHAPTER 2

Freethinker, with Misgivings

IN the first warm days of spring, when the horse-chestnut trees lining the avenues of the city were covered in buds, we children were obliged to take "a mouthful of sun" in the early hours of the afternoon on one of the avenues. The destination of these promenades for Paola and myself (our older brother and sister were already past this phase) was Corso Duca di Genova which enjoyed the advantage, over Corso Vittorio Emanuele where we lived, of being not a main artery for the city traffic but, rather, out of the way of public transport, used only by carriages drawn by broken-down old horses or by one of the rare automobiles operating during the first two decades of the century. Twin paths shaded by a double row of horse chestnuts flanked the central section. The dense foliage of April was followed by May blossom and the horse chestnuts themselves, which rained down onto the paths, some still encased in their spiky shells, and some not. We children were hard put to imagine how to use this large, stolid fruit, a mere caricature of the real edible chestnut. A row of benches under the trees provided nurses and governesses with a way of both keeping an eye on the children in their care and of getting their own share of the sun filtering through the greenery. Nurses were distinguished by being either "dry" or "wet." The unhappily named latter group, in addition to having to change nappies and keep the babies clean, had also to suckle them. They were strong young women, mostly from the Veneto region, and wore traditional dress—a velvet skirt and a cotton bodice with puffed sleeves, not unlike the picturesque costumes worn in their native villages on festive occasions. The governesses, who were in charge of the more grown-up children and shared the use of the benches with the nurses and retired folk and mingled with them, wore nothing special.

The governess who used to accompany us, and of whom I have the most vivid memory, was a certain Antonietta, known to us as Cincirla, a derogatory nickname we concocted and of which she was unaware; it was suggested to us by her extreme smallness and physical frailty, by the yellowish color of her face and the oblique slant of her small, extremely dark eyes.

These traits might have made one think she was of foreign origin and not from her real home, a small village in the interior of Sardinia. The nickname we coined, like all terms that become a part of a family lexicon, has engraved on my memory an array of sensations of the kind that, for better or for worse, last a lifetime in every one of us. The name Cincirla in my case evokes the recollection of those afternoon walks I abhorred. I disliked the contact of her tiny hand, which was strong and determined not to let go its grip on mine. I cared nothing for the "mouthful" of sun, even though grown-ups maintained that it was very good for one's health, and had no desire to show off my deficiencies in matters of sociability and sports. I was, in point of fact, the worst of all at skipping, hopscotch, ball, and badminton.

Our playmates of the same age varied from day to day according to the bench Cincirla chose for the customary two hours. They were, for the most part, little girls of bourgeois families living in the neighborhood, almost all of whom were Catholic and Piedmontese. As taught by their mothers, they would first of all ask for one's name, one's father's profession, and one's religion. My father was an engineer, a profession of considerable prestige to which there could be no objection. As for my religion, the first time I was asked I found myself somewhat at a loss because I had only vague ideas on the subject. Was I Jewish, Israelite, or devil knows what else? Since we went neither to church nor the synagogue (I knew of the existence of the latter not from firsthand experience but from a cousin of my age who went regularly), I wasn't certain what I should answer. My mother, to whom I posed the question, replied that it was better that I ask my father. He stroked my hair gently and answered with a seriousness that impressed me. "You children," he told me, meaning also my brother and sisters, "are freethinkers. When you reach twenty-one, you'll decide whether you wish to continue as before or whether you prefer to belong to the Jewish or the Catholic faith. But

19

don't worry about it. If you are asked, you should answer that you are a freethinker." This is what I did thenceforth, causing considerable perplexity in the inquirers who had never heard mention of such a religion.

Instead, it was Cincirla, who listened to what we said, who worried about it. She knew that our family did not go to church and was therefore not Catholic but Jewish. It had probably further motivated her, a bigot, to work for the family, in the hope of converting us—a merit that would have carried great weight at the Last Judgment. Naturally she would not have dared bring up the topic with my mother and even less so with my father who commanded tremendous respect, but it was not difficult for her to instruct us two little girls on the matter. Her attack began during one of those afternoon walks. "Do you know," she asked me, "what caused the earthquake in Messina?" Although I had heard of the terrible earthquake that in 1908, the year before I was born, destroyed that Sicilian city and stunned all Italy, I had, of course, no idea of its cause. Cincirla, however, was glad to inform me. "The day before the earthquake," she told me, "some Israelites went into a church in Messina and said in mockery before the image of our Lord Jesus, 'If you're the unknown God and not a fake, send us all a big earthquake.' Which is what happened no more than twenty-four hours later." (I should mention that the term *Israelite* sounded less crude at the time than the more widespread *Jew*, almost as if ameliorating the situation of all those—a small percentage of the population, all told—who professed the religion by misfortune of birth rather than by choice.) In spite of being very young, I wasn't convinced by the explanation, but it did upset me. "However," Cincirla went on to say, "you can save yourself from the curse hanging over people of your religion by coming to church with me and receiving holy water from the priest. In that way, you'll be redeemed, and after death you, too, will go to heaven." She had no doubt as to her own right to it. "And Mother and Father," I asked, "will they come with us?" "Unfortunately not," Cincirla answered, sighing. "Perhaps they'll only be able to join us when a dove that drinks once a year has dried up the sea." "If that's how it is," I answered without hesitation, "I'm staying with them." I didn't speak to my mother about our conversation lest she tell Father and irritate him; but from that day on, Cincirla knew the game was lost and gave up trying to convert me, abandoning me to my fate.

Freethinker, with Misgivings

The question of religion, however, remained a cause of great perplexity and discomfort during my childhood years. Father could not be considered an atheist or intolerant of those who had different ideas from his own in matters of religion; I remember the respect that he always showed to the clergy, both high and low. But he was certainly secular in the deepest sense of the word. He regretted not being able to enjoy the solace of belief; and on those rare occasions when he mentioned it, he gave the impression that he was referring to the Catholic and not to the Jewish faith. Though he did not go to temple (I remember accompanying him to Turin's majestic synagogue only once, on the occasion of my older sister's wedding) and entirely neglected religious rites and precepts, Father was by nature—and not out of a desire to comply with the laws—inclined to follow the norms codified in the Scriptures and ratified by tradition. This tendency, as his sisters used to recall, had been manifest since adolescence in the total dedication with which he faced all sorts of problems and in the absolute priority over wealth or power he gave to cultural, humanistic, or scientific interests. His energetic character, and the anger with which he reacted in his youth to what he considered bad behavior on the part of one of his numerous brothers and sisters, had prompted his sisters—who esteemed him greatly nonetheless—to nickname him "Damino the Terrible," using an affectionate diminutive of his name, Adamo.

My father, having lost his own father, a lawyer and man of letters, at the age of nine, had only a vague recollection of him. All his affection had gone to his father's brother who, in accordance with ancient Hebrew tradition, had married his widowed sister-in-law and fathered the last two of her eighteen children. This uncle-stepfather, a lawyer like his brother, known to us as "il Barba" (a Piedmontese term for "uncle") had been the dominant figure in my father's life. He remembered him with the most vivid tenderness and admiration for the limitless dedication he showed to the family and to the children of his brother, no less than to his own. Feeding such an extravagant number of hungry young mouths on a lawyer's small salary had exhausted him, and my father used to attribute his premature death to the financial problems he had to face, not just to keep the family but also to send the boys and two of the girls to the university. A faded daguerreotype of the period, which my father always carried, shows his uncle as he used to describe him to us: thoughtful, light-colored eyes, the features of a patriarch of Israel, a small pointed beard framing a rather wide, pleasant face. His expression was similar to my father's in seriousness but differed in the characteristic mildness and melancholy of his gaze, neither of these qualities appearing in photographs or in the extremely vivid recollection I have of my father. His light blue eyes, his well-shaped aquiline nose with nostrils that flared whenever he lost his temper (a warning sign which made me tremble as a child), an upper lip framed by a mustache and the lower one slightly protuberant, the severe expression that induced respect and fear not only in his children but in whomever approached him and felt his piercing scrutiny, all gave an impression of exceptional energy and determination. Father's personality had been formed in the cheerful and carefree environment of a large family where everyone had from early childhood a decided character, deriving more from their maternal than from their paternal inheritance. I know very little about the religious convictions of his parents. It seems that Barba wanted to make a rabbi of him but had immediately given up in the face of my father's clear aversion to the choice, though he had been well aware from childhood of being Jewish. Once a schoolmate who had been disparaging of Jews had been demolished by his young fists and advised not to try it again.

Turin under the Savoys, during the 1870s and the last two decades of the century, the years of his youth, had been increasingly tolerant compared with preceding decades; and all of Piedmont was pervaded by the liberal wind then blowing in a unified and hopeful Italy. In Rome, the city of the popes, the election of a Jewish mayor, who left behind a glowing memory, had lifted the spirits of Italy's meager Jewish population. News was constantly arriving, however, of pogroms in Russia and of Jews being isolated in ghettos, of the vexations they suffered because they belonged to a defenseless minority who refused to become assimilated and preferred suffering and humiliation to apostasy. Even though my father's secular spirit made him firmly discard the possibility of conversion, he still rebelled at the idea of perpetuating the memory of the persecutions undergone by his progenitors and most distant ancestors in order to keep alive a tradition that exerted no fascination upon him. Moreover, as a university student, he had established excellent relationships and lively friendships with Catholic classmates (two of his six sisters later found husbands among them); and he admired a Gospel message less ascetic and severe than that of the Old Testament. But his

Freethinker, with Misgivings

reluctance to engage in the rites and rituals of any religion, whether Catholic or Jewish, was such that, in the end he decided against giving any religious training to his children.

Thus, even before learning to read and write, much less to think, we became "freethinkers," which gave us certain advantages over our peers but rendered even more acute the sense of isolation from which our father had hoped to relieve us by this compromise. Traditional Jewish celebrations such as the Passover ritual, commemorating the exodus from Egypt and the deliverance from slavery, and the autumnal one of Yom Kippur dedicated to fasting for the expiation of sins, particularly highlighted our discomfort and are permanently printed on my memory. Not without some hesitation, and especially to please Mother who was not observant but more tied to tradition than he, Father went along with the solemn Passover ritual which for twenty centuries had been repeated with the same words and gestures by the descendants of the people of Israel scattered throughout the world by the Diaspora. The Seder, or ritual at dinner of the first two evenings of Passover, dates back to the departure from Egypt in the time of Moses. The Torah, or Law, prescribes that on these evenings one eat bitter herbs and a lamb sacrificed in the afternoon, and that the young be told of the events they are commemorating. This commemoration, read by the older or most authoritative member of the family, has become known as the Passover Haggadah, or Passover tale. The rite was celebrated by our family at the home of my mother's uncles, with her three bachelor brothers, her sister with her husband and their three children of our same age.

Every year I awaited the Passover celebration with a mixture of desire and anxiety. I liked the resplendent table set as prescribed by the solemn ceremony in which the mistress of the house—in this case our Venetian aunt, Anna, whom I adored—laid out beautiful plates, silverware, and glasses kept expressly for the occasion and personally prepared the dinner of exotic food which we children liked immensely. But the eating of dinner was preceded by the lengthy reading of the Haggadah, cause of my distress. After the food had been blessed by Uncle Teodoro, known to us as "Zio Doro," or by impressive Uncle Emanuele, there was the prescribed explanation of the fact that on that evening we would eat unleavened, instead of normal, bread, to remind the living that their fathers, fleeing from Egypt, had not had time to let it rise.

In orthodox families, all of the Haggadah was read in Hebrew; in my mother's observing but unorthodox one, it was read in Italian, and I and my brother and sisters would follow it attentively. Trouble would come when, after having picked up one of the unleavened loaves and taken the bitter herb from the Seder tray, the uncle in charge of the function continued the reading, giving thanks to our God and our fathers' God for having freed us from slavery. At this point, a slight flaring of my father's nostrils warned me that the critical moment was near. The thanksgiving, in fact, involved an enumeration of all the benefits for which we were indebted. These consisted not only in being permitted to flee from Egypt and reach the Promised Land, but also in God's punishment of the Egyptians with the ten plagues inflicted for the tribulations caused the Jews: lice, wild beasts, locusts, the death of the firstborn, and the others.

"What hatred!" my father would comment in a low deep voice, unheeding of my mother's barely murmured plea: "Damino, please!" and of the look of reproach Aunt Anna gave him. Imperturbable he would continue with what he wanted to say, turning to us, his children who, with bent heads were stealing glances at the others at the table: "Even if this resentment of theirs at tribulations suffered is human and has nothing to do with the persecutions they have always been subject to, it has nevertheless aggravated their suffering. I can't understand why, some five thousand years afterward, we have to take pleasure not so much in the exodus from Egypt and the end of slavery as in the fact that the Eternal Father punished our enemies with all these plagues."

A frozen silence and a brief pause in the reading of the Haggadah would follow. Uncle Doro, the oldest of the uncles and Aunt Anna's husband, who usually recited the prayer, loved Father dearly and, though he disapproved of his interruptions, believed there was no advantage in giving them greater weight by commenting in his turn. After a brief pause, he would resume reading.

When to our infinite relief, the reading was finished, we forgot the incident, at least temporarily, to enjoy the sumptuous dinner prepared by Aunt Anna, though in my confusion I felt that Father was right. The open conflict, however, between him and Aunt Anna, who had adopted and raised my mother from the age of two, was a cause of great pain. Father had great respect for her and was aware that his comments were out of place during the solemn ceremony, but the pressure he felt to warn us against supine acceptance of what was written in holy scripture in this specific case, and in general against the ordinances of authority, was too strong for him to renounce the chance of making freethinkers of us. I must add that he was far from being anti-Semitic. Though he did not follow its precepts, he fully recognized that he belonged to Jewish stock and was proud of the indomitable tenacity and cult of spiritual values he had admired in Barba as well as in other members of the family and in friends of his youth. His death in 1932 spared him from witnessing Hitler's rise to power and from being overcome like the rest of us by the whirlwind unleashed, certainly by a madman, but attributable also to servile obedience to the authority and power conferred upon him.

Yom Kippur was another occasion for head-on collision with our cousins, the children of my mother's sister, who atoned for their own and others' sins by fasting on that day for twenty-four hours. My father maintained that as freethinkers we were exempt from any such obligation. Yom Kippur occurred at the beginning of the autumn, in the month we used to spend with our cousins in a villa in the Asti hills in Piedmont. The villa was surrounded by a large park, a vineyard, and an orchard full of ripe fruit, all of which were for us a source of great joy. It was therefore not difficult for us to obey our father's orders, while ignoring the scornful glances of our cousins as they walked up and down the park counting the hours. The fast would end when three stars had appeared in the clear sky of summer's end. We freethinkers with full stomachs tried to help our cousins and show our zeal by scrutinizing the sky and rejoicing whenever we thought we had spotted the three stars. The cousins, however, took no heed of our claims, maintaining that only the ones they saw could be counted. They immediately reported to Aunt Anna who, anxious to appease their great hunger, had already prepared the *bruscadella* which was customarily served when the fast ended. The *bruscadella,* for which we children were gluttons, was nothing more than slices of toasted bread flavored with cinnamon and other fragrant spices and soaked in sweet wine. The famished cousins would throw themselves on soupdishes filled with it: we freethinkers were allowed to serve ourselves only after they had satisfied their hunger. Since they had expiated not only their own sins but ours as well, we too were in some measure entitled to it.

CHAPTER 3

Two X Chromosomes in a Victorian Climate

De tous les Préjugez on n'en a point remarqué de plus propre à ce dessein que celuy qu'on a communément sur l'Inégalité des deux Sexes. . . . Aprés avoir examiné cette Opinion suivant la regle de verité, qui est de n'admettre rien pour vray qui ne soit appuyé sur des idées claires et distinctes; d'un costé elle a paru fausse, et fondéé sur un Préjugé, et sur une Tradition populaire et de l'autre on a trové que les deux Sexes sont égaux: c'est a dire, que les Femmes sont aussi Nobles, aussi parfaites et aussi capables que les hommes. Cela ne peut estre établi qu'en refutant deux sortes d'Adversires, le Vulgaire, et presque tous les Savants.*

—FRANÇOIS POULAIN DE LA BARRE,
De L'Egalité Des Deux Sexes, 1673

AT THE AGE OF FOUR, I was not acquainted with the fact that the quite evident physical difference and the presumed intellectual one between male and female individuals of our species are due to the possession of an X and a Y chromosome, on the one hand, or of two X ones, on the other.

Possessing two X chromosomes, it was my fate to grow to womanhood in a period when the natural intellectual faculties of an individ-

---

*Among all the prejudices we have come across, none has seemed more pertinent to such a design than the one commonly held on the inequality of the two sexes. . . . After having examined this opinion according to the rule of truth, which is that of not accepting anything as true which is not supported by clear and distinct ideas; on the one hand, it appeared false and founded upon a prejudice, and upon a popular tradition and furthermore we found that the two sexes are equal: that is to say, that women are as noble, as perfect and as capable as men. Such a thing cannot be established except by refuting two sorts of adversaries, the vulgar person, and almost all learned ones.

ual possessing an X and a Y—that is, a man—were reinforced rather than repressed. In the same way, in archaic societies and in many that still flourish today, the fact of having inherited genes from parents of high or low social rank definitively marks the destiny of the newborn infant. During the last century and the first decades of our own, in the more "advanced" societies (so termed according to the belief that industrialization equals progress), a woman faced an almost insurmountable barrier if she wanted higher education and to realize her gifts. Though the Victorian era passed away at the beginning of the twentieth century, its enduring influence on the kind of education given to young members of the two sexes continued to determine their roles.

The Victorian ideal was given expression by John Ruskin, the foremost art critic of the nineteenth century and one of the most influential personalities not only in the arts but also in English society in the second half of the century. In *Sesame and Lilies,* first published in 1865, he expounded his credo on the education of women:

> She must be . . . wise, not for self-development, but for self-renunciation.
> . . . A man ought to know any language or science he learns, thoroughly,
> while a woman ought to know the same language, or science, only so far
> as may enable her to sympathize in her husband's pleasures, and in those
> of his best friends. . . . She is to be taught somewhat to understand the
> nothingness of the proportion which that little world in which she lives and
> loves, bears to the world in which God lives and loves . . .

The physiological complementarity of the sexes was, according to Ruskin, clearly evident in the roles nature had assigned them:

> The man's power is active, progressive, defensive. He is eminently the
> doer, the creator, the discoverer, the defender. His intellect is for specula-
> tion and invention; his energy for adventure, for war, and for conquest.
> . . . But the woman's power is for rule, not for battle,—and her intellect is
> not for invention or creation, but for sweet ordering arrangement, and
> decision.

After a long-winded investigation of the roles assigned to them, Ruskin concluded that woman is better than man. John Stuart Mill, in *The Subjection of Women,* published in 1869 as a reply to Ruskin, stated of women: "They are declared to be better than men; an empty compliment, which must provoke a bitter smile from every woman of spirit, since there is no other situation in life in which it is the established order, and considered quite natural and suitable, that the better should obey the worse."

Ignorance of historic precedents could not protect me from the consequences of having been born at a time when the influence of the Victorian era and its pervasive romanticism were still very much alive in my parents' generation which found in it strong backing for patriarchal rule. Our little family group was an exemplary model of the way such ideologies are applied. Mother and Father both came from Sephardic families which had moved respectively from Asti and Casale Monferrato, two towns of some importance in Piedmont, to settle in Turin at the turn of the century. They differed in regard to religion only in that Father's family was more overtly secular in its attitudes, whereas Mother's, as I have mentioned, was observing, though unorthodox, and more tied to tradition. One of my mother's three brothers became a well-known ophthalmologist in Turin; another, Uncle Manno, whom I liked best because of his resemblance to Mother (both being tall, and blond enough to be mistaken for Anglo-Saxons), was mild-mannered and had a lively intelligence. He had taken a degree in chemistry but preferred to dedicate himself to problems of farming and the administration of a little town in the Asti hills where we had our summer home. He became its mayor and benefactor, bringing in electricity and the telephone and building a school—my brother's architectural debut. At Uncle's request, Gino also made a life-size statue of Defendente, the town's patron saint, which stood on a plinth in front of a church at the top of a hill, the turning point of our Sunday walks. I was very impressed with it, as with everything that Gino did.

Actually, I now think this statue must have been completely lacking in artistic merit, but it entirely satisfied the aesthetic requirements of the saint's devotees. The town was so grateful to my uncle for his dedicated efforts and the improvements he made during his long term of office that many years later, when the Fascist regime aligned itself with the Nazis, the townspeople went against orders and named their most important street after him.

Mother and Father met, as I discovered recently from a minuscule diary I found among the family papers, on 10 January 1901; she was twenty-one and he was thirty-three. In the diary she had written—in the same, unmistakable, tiny handwriting of her last letters to me

28

Two X Chromosomes in a Victorian Climate

before her death at the age of eighty-four—"10 January 1901: an unforgettable day." The sentence is underlined and followed by a quotation in brackets from Paolo Mantegazza, a romantic author who was in vogue at the time: "The handshake is, for friendship, the last letter of the alphabet, for love, the first." The love born from that handshake lasted till Father's death, thirty-one years later. Mother used to recall with amusement that, immediately on meeting, he had revealed in a very worried fashion what he considered a serious physical impediment: he had a false tooth. He made no mention of his past attachments, which cannot have left a deep or lasting effect upon him. He had a liking for his beautiful flirtatious cousins Celeste and Ottavia, for whom Mother nourished, perhaps unconscious of the reason, a strong dislike from her first encounter with them. Given her sweet disposition—to my recollection, Mother never thought or said ill of anyone—I was rather surprised as a young girl to hear her call these cousins "conceited gossips." Father would smile and stay silent.

I find it hard to remember when it was that I first became aware of the different roles that Father and Mother played in the running of the family, and of the enormous extent to which my father's will prevailed, unchallenged in decisions large and small concerning us. I believe that a certain vague awareness dates back to an occasion when I was between two and three years old. Paola's bed and mine were aligned next to each other along one wall of the room, parallel to our parents' large double bed. Waking up in the middle of the night, I noticed that Mother's hair, which in the day was coiled around her head, was now let down and spread out on the pillow, giving her face an entirely different aspect. I immediately woke Paola to share my discovery with her, whispering so as not to disturb our parents: "Pa, Pa (my name for her has always been Pa and not Paola), Mother has a *ciuffetto* (forelock)." Though not the correct term, it served my purpose. Father had waked up as well and murmured something to Mother. The next morning, very gently and without making any reference to my nocturnal observation, he informed us that from that night on we would sleep in a different room. This rather insignificant episode has remained so vivid in my memory, however, that I recall all the details of the bed and my own amazement at observing Mother's hair arranged in such a different way from what I was accustomed to.

Dating back to that same period is another small episode that made me aware that every decision concerning us was up to him and not Mother. The day in question we were wearing new straw hats, wide brimmed, and set with ribbons and little flowers which seemed immensely fine to me. Upon returning home from our walk, proud of our purchase, we greeted Father who had just come home from the office. He looked at us in amazement and with obvious disappointment. "These hats," he said, turning to Mother, "are in terrible taste. They mustn't wear them." Mother timidly objected that we liked them a great deal. Impatiently Father answered that he never wanted to see them again, and his will was done. An endless number of similar episodes during our pre-school and first school years concerning Paola and myself, as well as our brother and sister, convinced me that in spite of the fact that we only saw him at lunch and dinner and that he was often away in Bari for weeks on end managing the big plant he had built, it was he who controlled our lives, even in small details.

The severity and the piercing quality of his gaze, the slight flaring of the nostrils preceding his brief but violent outbursts of anger, the imperious voice which contrasted so sharply with the sweetness of my mother's, not only pointed to differences in their personalities but offered me the first tangible evidence of just how different were the roles of men and women in the society of the day.

My father very much loved the opera and the theater which were then deeply romantic. The Turinese public's favorite actor at the time was the famous Ermete Zacconi, an imposing figure with a deep, vibrant voice. One of his best-loved performances occurred in a play called *Il Padrone delle Ferriere (The Foundry Owner)*. Zacconi naturally played the lead, a captain of industry who acquires riches and success by building Italy's first railways at the end of the last century. No longer young, he marries, or—as it was customary to say at the time—"wins the hand of," a young and beautiful woman with whom he has fallen madly in love. She, however, does not requite his love as he had expected, and a stormy period of unpleasantness and quarrels follows the marriage. After one particularly violent clash, the foundry owner confronts her as follows: "You love me," in a low, hoarse voice; "I adore you," gradually increasing volume and register; "but I'll break you." This last phrase—Zacconi's tremolo rising to a crescendo, before he fell back into an armchair, pale and overcome with emotion—would drive the audience to wild applause. Father

30

shared the general admiration for Zacconi's expressive powers and would repeat the famous sequence to Mother. She would smile since there wasn't the slightest resemblance between the couple in the play and Father and herself. We little children (I must have been seven or eight) kept quiet, but I remember the deep resentment I harbored for Zacconi and my hatred for the foundry owner who considered himself the proprietor not only of the foundries but also of his wife, while the audience approved.

As for opera, Father would, in the morning while shaving—a lengthy operation with the razors of those days—sing his favorite arias. At the top of the list came Verdi's *Il Trovatore* and *Aida, Tosca, La Bohème,* and *La Traviata.* His beautiful baritone voice resounded throughout the apartment, and his singing made us happy because it was a sign he was in high spirits. But I couldn't listen without irritation to the piece from *La Traviata* in which the father reproaches his son for having insulted his mistress, believing her corrupt (whereas in reality she is not). In the person of Papà Bermont, Father would shave and sing the famous phrase, "Di sprezzo degno se stesso rende, chi, pur nell'ira, la donna offende" (he who offends a woman—even if only in a fit of rage—makes himself worthy of scorn). Though chivalrous in the Ruskin way, this attitude toward women, whether innocent or guilty, provided further evidence of a male superiority I could not accept. Father never suggested, however, that he shared the then common opinion that the superiority of men derived from greater intellectual capacity rather than greater physical strength. I remember the joy I felt when he said the most intelligent of his numerous brothers and sisters was undoubtedly his sister Costanza, known to all as Tina. Only slightly younger than he, she had shown since early childhood an extraordinary talent for music and mathematics. Many years later, when the much admired Aunt Tina was over sixty, she turned to sculpture with considerable success.

It was not my father, therefore, but my male cousins on my mother's side, with whom we used to go on summer vacations, who would amuse themselves by tormenting me. Knowing how sensitive I was on the issue, they boasted of how immensely superior in intellect men were to women. When had there been a Newton, an Einstein, a Bach, a Beethoven, a Michelangelo, a Leonardo, or a Raphael (and the list of names would continue for hours) of the female sex? My limited notions, coming at the time not so much from school as from a passionate reading of a children's encyclopedia, confirmed the assertions of Giorgio and Paolo, who were, respectively, five and one years older than I. My only defense, and all I could boast, were Sappho, Vittoria Colonna, and Gaspara Stampa, whose poems had earned them a place in history. Giorgio and Paolo would not only point out how few they were compared with the number of "real poets," as they called them, but also claimed that in any case it was not necessary to be intelligent to write poetry. I was grateful to Gino for not lending his voice to their cause. Taken up as he was by his passion for sculpture, he worked from morning to night with his plasticine and paid no heed to such problems.

CHAPTER 4

The Choice

THE ELEMENTARY SCHOOL was situated quite near where we lived; and, in accord with the principles in force at the time, girls were separated from boys in different classes. The vast majority of the children who attended were from working- or lower-middle-class families; a small percentage had fathers who were professionals, or came of wealthy parents. There were even then private schools, mostly run by religious orders. The choice of school had been dictated by its nearness and by the wish, shared by Father and Mother, that our schoolmates should come from all social strata and not just from the privileged classes.

The four years passed in that school have remained with me as a pleasant memory. The schoolmistress, the third of three sisters all of whom had devoted themselves to elementary teaching, carried out, with both skill and a high sense of duty, what was for them a mission. From the very first day, I loved my teacher for her stern manner, her tall slender figure, and her sweet and melancholy expression. Like her sisters, she was inflamed with love of country, particularly lively in those days, for while we were in first grade, Italy entered the First World War alongside the French and English. This passion of hers infected me, and I declared, to her satisfaction and to the lesser enthusiasm of Father and Mother, that our king and queen held first place in my love—followed, of course, by my parents. These patriotic sentiments were enhanced by the rare visits of our schoolmistress's elder sister—tall like her, slender and austere, a Red Cross nurse who, they said, was sometimes on duty at the front. Her white uniform with its large red cross, and her devotion to the cause and veneration for the duchess of Aosta, the royal princess who was patron of the nurses, filled me with enthusiasm. Such was my admiration for her that I hoped the war would last long enough that I, too, could join the Red Cross nurses and take part in some heroic action on the battlefield. Happily, this hope was not realized. After the initially shocking defeat at the hands of the Austrians at Caporetto in October 1917, came decisive victories and peace. And in the third grade, we read with pride and emotion the last communiqué—learned by heart, naturally—of the supreme commander of the Italian forces, General Armando Diaz: "The enemy are in rout and, with our victorious troops in hot pursuit, are fleeing hopelessly and in disorder back through the valleys from which they came down with such arrogant confidence four years ago."

When we had gotten through elementary school with flying colors, a middle school had to be chosen—a choice that would determine what followed: university, artistic or technical training, or teacher training for the elementary school. It was an important decision for the boys, but of little or no importance for the girls, since it was taken for granted that awaiting them was the career of being a good wife and mother. Despite the fact that all three of us girls had demonstrated outstanding aptitude for study, our father decided that we should attend middle school and then the girls' high school—from which, in those days, there was no possibility of going on to the university. The girls' high school differed from the boys' not so much in the teaching of language and literature as in the training in mathematics and the so-called exact sciences, though even in the boys' school this was limited to the teaching of basic notions. Our father's decision to channel us toward the girls' high school not only accorded with that of the parents of friends and acquaintances but also derived from his knowledge that two of his sisters, who had—altogether unusually for their time—earned doctoral degrees in literature and in mathematics, had then found great difficulty both in going on with their studies and in reconciling them with marital and maternal obligations. Today, almost a century later, these same difficulties still exist, and the problem is far from being resolved. In the case of my eldest sister, Nina, who had shown a strong bent for mathematics, the decision taken by Father—whom, like Paola, she adored—did not create grave problems. She had a decided preference for the arts and was determined to become a writer, for which no degree would be necessary. Though, like me, Paola would have much preferred to attend the boys' lyceum, the exceptional artistic aptitude she had

34

shown since childhood rendered Father's decision less difficult for her as well. When she was seventeen, Paola finished high school with excellent grades and joined the atelier of Felice Casorati, artist of European fame. From those early years up to the present, she has dedicated herself with the liveliest perseverance and enthusiasm to following her chosen career as a painter.

Of the three sisters, the one who found herself in greatest difficulty was myself. To the lack of artistic gifts, highlighted by the contrast with Gino and Paola, was to be added the conviction that I possessed no other particular talent. The biological sciences were not part of our high school's curriculum. To me, a bat was no different from a bird, both being flying creatures; nor had I any clear ideas about the difference between the various inhabitants of fresh and salt water, from crustaceans to fish or whales. I held the belief, but nourished well-founded doubts, that I had a bent for philosophy, and would have enrolled in that faculty if the girls' high school had allowed entry to a university. Unable to direct myself as Paola had, I drifted along in the dark and tried to fill the emptiness of my days by reading Selma Lagerlöf—more owing to Nina's encouragement than out of spontaneous interest—and other books she suggested. A natural aversion to sports, and an even stronger one to establishing contacts with girls of my own age, accentuated my profound sense of isolation, which derived in part also from timidity and my total lack of inclination to approach young men of my age or older in the expectation of meeting a future life companion. My experience in childhood and adolescence of the subordinate role played by the female in a society run entirely by men had convinced me that I was not cut out to be a wife. Babies did not attract me, and I was altogether without the maternal sense so highly developed in small and adolescent girls. This feeling was strong instead in Nina, who was ready to give up her aspirations to be a writer when a meeting—arranged by the parents, as happened in those days—with a distant cousin, an honest businessman with a fine reputation, led to marriage; the children, who followed at an unusually rapid rate (three babies, including twins, in eleven months), absorbed all her time. Her engagement marked the end of our ambition to emulate the Brontë sisters who had by then replaced Lagerlöf: the spell of those frozen Swedish winters had given way to that of the wild empty spaces of the parish where the Brontës' father exercised his ministry and the three sisters, unlike their shiftless brother, went through the literary apprenticeship from which Emily's masterpiece was born.

While Nina was taken up with preparations for her marriage, and Paola was working with feverish ardor both in the studio and at home, I was asking myself how I was going to escape from the blind alley I seemed to be in. A tragic occurrence put Ariadne's thread into my hand. The three female figures whom from my early childhood I had seen, each in a different way, as my guardian angels and loved with immense affection were my mother, Aunt Anna, and Giovanna. Giovanna, who was two or three years younger than Mother, had joined us before Paola and I were born. She came from a little village in Piedmont, where as a child, the third of five sisters, she had undergone privation and suffering. Her father was a peasant, tough and too quick to use his hands, who at four in the morning, winter and summer, would toss them out of their beds to tend the few animals they possessed and learn to earn their bread. Her mother had died shortly after her last pregnancy. When Giovanna came to work for us as a governess, she immediately formed an attachment to Mother and felt a sense of veneration for Father, who treated her with lively sympathy and great gentleness. As for us, we were her children in some measure. I remember her night watches when we had the inevitable childhood rashes or a slight temperature. In spite of the difference of religion—naturally she was a Catholic—she was not a fanatic like her stern oldest sister, Caterina, in service with the parish priest of her village; nor was Giovanna cheerful and mocking, like the youngest sister, Bettina, who also came to our house as governess. Neither Giovanna nor Bettina ever touched on the question of religion, as Cincirla had done, and even less did they try to convert us.

During the months that I was struggling with my doubts, I had noticed Giovanna's pallor—without worrying about it, however, so taken up was I by my problems. It did not, in fact, seem substantially different to me than usual. Mother, on the other hand, was worried and begged her to see our family doctor. I remember the brief note he gave her in a sealed envelope to deliver to Mother, and which we read together. "Giovanna Bruatto," he wrote, "whom I have examined, shows all the symptoms of a serious illness. I fear it may be cancer of the stomach. She should be admitted to hospital immediately."

She went in the following day, and the grim diagnosis was

confirmed. She underwent an emergency operation while we waited outside the door. In a state of growing anxiety, we heard from the surgeon the fearful news: there was nothing to do. The cancer, which was blocking most of the gastric tract, had spread and invaded the surrounding tissue. When she returned home to us, still unaware of the gravity of her illness, Giovanna found comfort in our tenderness and concern to save her any chore. I remember her thin and emaciated, sitting on a kitchen chair outlined against the gray sky of autumn: she contemplated it with her hands crossed in her lap. It was on that day that my decision took form, and I felt that I would be able to persuade Father. I would take up my studies again, and go into medicine. I told Giovanna of my decision, adding ingenuously that under my care she would certainly get well. After all, she was young, hardly forty-five, and we would spend together the many years remaining to us. I remember she squeezed my hand and smiled with immense sadness. *"Masnà,"* she said to me, using the Piedmontese word for "child" which she always used with Paola and me, "when you are a doctor I shall have been in the Elysian Fields for years." I was startled by this mention of the Elysian Fields. Without any education and a practicing Catholic as she was, why not speak of Paradise rather than of that mysterious and mythical abode of happy souls after death? Where had she learned of their unlikely existence? Certainly not in church. I believe that in her modesty she never would, and never could, have hinted at the possibility of going to Paradise. The Elysian Fields, which she might perhaps have heard us mention when studying mythology, represented a kind of compromise between what she hoped for and what she felt lay close at hand: the passage over into the beyond. A few weeks later, she asked Father and Mother to let her return home, to die close to her unmarried sisters who still lived in the little farmhouse where they had been born and seen the sun rise on summer days and shivered on winter ones. We accompanied her to Rivarossa, the little village, about an hour from Turin, which she had often mentioned when telling me of her childhood. The news that reached us in the following days confirmed the nearness of the end. I saw her for the last time one cold November day in 1930.

We went to see her—Paola, Mother, and Aunt Anna. Carlo, my aunt's chauffeur, drove the car. Since their first meeting many years previously, Giovanna had conceived for him a passion bordering on adoration. Carlo was neither handsome nor particularly attractive, but his virility and other qualities she saw in him had attracted her. For the first time in her almost nunnish life, she had felt an irresistible impulse toward the other sex—but disastrously, for Carlo had decided to marry Aunt Anna's cook whom Giovanna detested.

Now we drew near her bed. Was that Giovanna? The face with its ghostly pallor was no more than a skull with staring eyes veiling death. One could see the skeletal body under the covers. I stroked the hand that had often caressed and comforted me, and called to her, but she gave no sign of feeling my touch. Carlo, standing next to us, spoke to her in a low voice, in Piedmontese: *"Couragi Giuaninna, andrà mej"* ("Courage, Giovannina, things will get better"). Her face remained motionless, but a tear on her lashes showed she had heard the beloved voice. A few days later, we went to her funeral.

I told Mother of my decision to study medicine. She encouraged me to speak to Father. I believe that when, the following day, I timidly asked if I might have a word with him after dinner, he had already been informed by Mother. He said I might have it there and then. I began in a roundabout way, telling him that, since I had no vocation for married life or for having babies, I would like to go back to studying. He listened, looking at me with that serious and penetrating gaze of his that caused me such trepidation, and asked whether I knew what I wanted to do. I told him how much Giovanna's death had shaken me and how I was convinced that the profession I wanted to follow was that of medical doctor. Even if only in a vague way, I had been drawn to it from my childhood during the war, when I had wanted to be a Red Cross nurse in the belief that being a doctor was something to which I could not aspire. Father had been no less distressed than we at the loss of our Giovanna, whom he had often gone with us to see while she lay in the hospital. He objected that it was a long and difficult course of study, unsuitable for a woman. Since I had finished school three years previously, it would not be easy to take it up again. I assured him that I was not afraid of that. With the help of a tutor, I would study privately. "If this is really what you want," he replied, "then I won't stand in your way, even if I'm very doubtful about your choice."

At that time I was barely twenty. I knew that I would have to overcome great difficulties in taking up my studies again, especially subjects like Greek and Latin, in which I had no training. My knowledge of mathematics was rudimentary as well. Happy with his consent, however unenthusiastic it was, I asked my cousin Eugenia, who was a year and a half younger than I and had had similar schooling, whether she wanted to join me. Having lost her father when she was nine, she was dependent only on her mother's benevolent authority, and stoically bore the bullying and mockery of her elder brothers. As a child she had been their plaything, and one of their tricks had been to teach her to ceremoniously remove her little girl's hat—which in those times was very large, decked with velvet ribbons, little birds and flowers, and held in place by a chinstrap—just as they did with their little berets whenever they encountered a distinguished friend of their parents in the street. When she was five or six years old, they set her simple mathematical problems, such as calculating the number of chestnuts she could get with thirty *soldi* if ten bought thirty. Her prompt but not very correct reply that she would get "a little bagful" became part of the family lore. It was expected that both boys would have a brilliant future, but nobody was interested in what hers would be, so rooted was the conviction that it would be that of all the daughters of rich parents in Turin's *haute bourgeoisie:* a future divided between family obligations and receptions. In her case, however, the unthinking conviction proved quite wrong.

Upon finishing school, Eugenia had made friends with cousins on her father's side, who were more inclined for social life and going out than were Paola and I. Nevertheless, Eugenia enthusiastically approved my project. To act on it, all that we required was the help of two teachers: one for Latin and Greek, and the other for mathematics. The first to go along immediately with our request was Professor Lobetti-Bodoni, a friend of the family who was known in Turin for his extraordinary knowledge of, and ability to teach, Latin and Greek. The other teacher was the diminutive and famous (at least in Turinese circles) Professor Guido Ascoli, from the best-regarded scientific high school in town. As for philosophy, literature, and history, we would be able to manage those on our own.

I threw myself headlong into this program which had become the very reason of my existence; more frolicsome and gay than I, Eugenia did not give up her social life, though she studied with great seriousness. We had promised ourselves to complete our preparation for the examination in eight months, starting in February, and to present ourselves as "external" candidates (those who studied at home rather than attending classes) at the autumn, rather than the summer, session. To get through the Latin program, and Greek which we had never touched, we decided to spend the summer in the mountain village where Lobetti-Bodoni went to stay. I used to get up at four in the morning; and Eugenia, who shared the room, would climb groaning out of bed and, more asleep than awake, sigh her way with me through the lesson Lobetti-Bodoni had given to us the day before. When we showed up for the autumn examinations with candidates who were repeating them and with other external examinees, I was extremely worried, while Eugenia was as gay and witty as ever. I remember the opening sentence of the Latin text we had to translate: *"Qui ad solem venit,"* claimed some author or another, *"licet non ad hoc venit, colorabitur."* The statement pleased me very much as apparently referring to my own case. As anyone exposed to sunlight gets tanned, willingly or not, by the sun's rays, so I was taking in, without being entirely aware of it, the joy of deepening knowledge. The similarity, in reality, was not a strong one, but it suited me and helped me translate the entire passage in record time. A blonde girl sitting at the desk behind me kept wailing, "Rita, dear Rita"—whispering my name though our acquaintance had begun only half an hour before—"what the devil is this wretched author saying?" In a whisper I set her on the right track, and she was very grateful. The other exams went well, except geography. I wasn't able to name the currents in the Pacific Gulf or say why they formed. Mutely, I watched while the woman examiner wrote down her comment: "The candidate shows she has no idea what the Pacific Gulf currents are." This serious lacuna was a nightmare to me in the days that followed, while I waited, with anxiety increasing hourly, for the outcome of the exams. I shall never forget the telephone call from Lobetti-Bodoni, his voice breaking with emotion, for he had been awaiting the verdict with trepidation almost as great as mine. I don't think I have ever again felt the joy I felt when, before the news had yet been posted, he told me in exultation: "Signorina Rirì, you've both passed!" I had come out at the head of the list of candidates, despite the business of the currents. I immediately let Eugenia, who had been waiting for the results with cheery nonchalance, know. Whereas I was determined to enroll in medicine, she was more inclined to sign on for mathematics; however, I persuaded her to go for medicine. I think that afterward she was happy with her choice.

The Choice

As for me, I was grateful to my father for having kept me back from the university for so long. If I had studied classics after middle school, as had been my wish, I would certainly have enrolled in philosophy, a subject demanding strong powers of logical thought which, to my misfortune, I lack. But, I wonder, how many professional philosophers are really gifted with them?

CHAPTER 5

The Death of My Father

THE YEAR 1932 was to leave an indelible mark upon my life, a mark that will be erased only when, in the austere words of Ecclesiastes, dust shall return to dust and the spirit to God. At the end of May—as I was completing my second year of medical school—Father was struck for the first time with an episode of cerebral ischemia which, though only brief, worried him. We, in turn, were alarmed, never having heard him mention, much less worry about, his health. Coming home late one afternoon, I was surprised to find him sitting on the dining-room balcony, so lost in thought as not to notice my presence or my greeting. I questioned Mother about this unusual behavior. She told me that a few hours previously, in the factory with his workers, he had noticed that he was having difficulty expressing himself and especially finding the words he wanted to say. His rapid return to normality in the following hours, and the fact that no similar episode occurred again in subsequent days and weeks, reassured us—but not Father, who well realized its sinister meaning. He continued to get up very early, as was always his habit, and to go straight to his factory, situated outside of town at a considerable distance from the center where we lived. With our automobile—a Storero, a clumsy old vehicle built by a local plant which, overwhelmed by competition from Fiat, had closed down shortly after starting production—he could have reached it in about half an hour. Instead, he preferred to use public transport, giving the excuse that the cold morning air did him good. Mother, who was persuaded he did it to economize, tried in vain to persuade him to spare himself the extra fatigue, especially after that episode and the ever more frequent anginal pains which he

The Death of My Father

seemed to ignore. When he was out strolling with us and they would halt him in his tracks, he would smile and say, "Don't worry, they're little pains of no importance." He did not have the time to think about them or to see a doctor, driven as he was by his passion for his machinery and by the urgent desire to realize his dream as quickly as possible.

The dream was to rebuild in Turin, even if on a smaller scale, the ice factory and the big alcohol distillery he had built in Bari, where he had first set himself up in industry a few years after graduating in engineering at the Turin Polytechnic when only twenty-one. He had started with the ice factory and taken his young wife with him after the honeymoon. Those difficult years he, nonetheless, remembered with joy.

At first he had to face the sharp hostility of the people who supplied snow to the local butchers, a curious and not easily definable category of tradesman who in the winter brought down snow by the cartful from the nearby mountains and stored it in underground tunnels not far from the city. In these tunnels the snow lasted till the summer, when it was sold to the butchers who used it to keep their meat fresh. The production of ice in my father's small but efficient factory was dangerous competition. In the mild climate of Bari, on the Adriatic coast in the south of Italy, the locals were not used to seeing ice; they knew only the odd light dusting of snow which melted in a few hours. Never before had they seen ice, and certainly never ice produced artificially in solid rectangular blocks, transparent as glass. The owners of the stored snow, alarmed at the idea of losing their customers, spread the rumor that the young Jewish engineer who had landed on them from the north was an emissary of the devil and that his products would bore holes in Christian guts. Glass it looked like and glass it was: so they spoke of the blocks like gigantic ingots which came out of his factory.

Mother remembered that there had been threats to kill him unless he stopped production and returned to the north whence he had come. These threats, while they made him laugh, terrified her. Trembling at the possibility of his being attacked and killed, she would wait up until two in the morning for him to return to the little house where they lived. A splendid daguerreotype of the period—June 1902—untarnished by the years, depicts her on the balcony where she spent her long vigils, with her firstborn, Gino, in her arms. Her excep-

43

tionally beautiful face, framed by blond hair gathered at the nape, bends, with a serious thoughtful expression, over his equally beautiful one, which is frowning at his own two minuscule big toes sticking out from under Mother's arm. The striped blouse covering her young, slender figure and her erect bearing as she leans on the railing evoke not a madonna but a figure from Dante Gabriel Rossetti's paintings. So she must have appeared to whomever saw her then, and so she remained until old age, though her hair had become silvery white and the still slender body had lost the firmness of youth.

After this first period, Father's enemies, who were not really anti-Semitic (a sentiment hardly known in southern Italy in those days) but attacked him to defend their business, were mollified by his cordiality and generosity and became his best friends. But making ice was not nearly enough to absorb Father's enormous energies, and he built a much larger factory near the first for the distillation of alcohol from carobs. The factory had grown quickly and gave work to a thousand people, as Father used to tell us proudly when he spoke of that time. After the war, things began to go wrong. A trainful of carobs headed for Bari, from I don't know which Far East country, caught fire (it was later shown to be arson), and most of the damages were never paid. After that, there were the so-called white strikes, the occupation of factories by the workers, which pained him all the more because he loved them deeply and had always been close to them. The strikes, which stopped production for months on end, and the crisis that hit that sector of industry, transformed the plant's former profits into growing losses. For the first time, Father had to resort to IOUs. Though I was quite young at the time, I remember the anxiety I felt in the air and in the low-voiced conversations between my mother and father at the approach of due dates. Bankruptcy, a thing my father considered too shameful to outlive, would have been inevitable had not my mother's three brothers—who were both well off financially and unmarried, and had therefore no family obligations—intervened to help. But it is all too well known that nothing is ever given—except through intimate love and affection—that does not have to be paid for in one way or another. In this case, these brothers—one a businessman, and the others respected professionals, good-natured men who had always been on excellent terms with Father—met his obligations though in fact disapproving of his unsolicited and unwelcome remarks about the exodus of the Jews from

44

Egypt. While they signed the checks necessary to pay the IOUs, they could not help commenting that none of this would have happened if Father had been a better and wiser administrator. Mother, who was by nature mild mannered and extremely sweet, intervened with unexpected energy, pointing out that neither the burning of the train nor the white strikes nor the general economic crisis could be blamed on "Damino." From then on, the bond between them became even stronger, and she never left him for a moment, going with him on every trip to Bari and constantly sharing his troubles.

It was in this "defeated" frame of mind that Father decided to rehabilitate himself—vis-à-vis his brothers-in-law but especially in our eyes—by rebuilding in Turin the ice factory and a new distillery, the one in Bari having closed down. He gave himself to the project with all his characteristic energy, despite the fact that he was now over fifty and exhausted by the battle to avoid bankruptcy. It was only after his death that we found out from his closest and most devoted collaborators, in large part workmen, how often during the days following the ill-omened episode of May 1932, he had slumped in distress, pressing a hand to his heart and asking them not to say anything about it to Mother.

On 30 July he had a severe attack of angina during the night, and Mother, who was constantly alert, administered the nitroglycerine the doctor had told her to keep on hand. I was then in my second year of medicine, and before giving it to him, Mother had come to me. With her and Paola, I anxiously waited to see whether the drug would relieve the pain. When it did, Father told us with a smile to keep calm and go back to bed. At six o'clock in the morning, I peeped into their room to see whether he was asleep. He was not, but he had gotten up and was sitting in an armchair looking rested and with a cheerful expression on his face. "I'm feeling fine," he assured me. "It's all gone."

An even more serious attack occurred in the night between 31 July and 1 August. On the morning of 1 August, the attacks of angina increased in frequency and were accompanied by an acute pain in the right side of the body. The family doctor acknowledged that the case was serious and arranged for one of his assistants to be always on hand, installed, unbeknownst to Father, in a room near his, ready for any eventuality. What turned out to be the final attack began at seven in the evening, his favorite time of day, which he loved to spend in contemplation of the swallows in flight. Often after the ischemic attack, I had heard him murmuring to himself: "And afterward?"

On the evening of 1 August, as on the previous day, he suffered from a lacerating pain on the right of the abdomen. The doctor on duty and the one living in the building, who rushed to Father's bedside, diagnosed heart failure and subsequent deficiency in the liver's circulatory system. Convulsions and loss of consciousness ensued, interrupted by brief periods of recovery. During the latter, in the silence of the room, surrounded by the doctors and by us—Gino, Nina, Paola, and myself (we had begged Mother to go and rest, and she had heeded our pleas)—he recited stanzas from *La Divina Commedia* in his deep voice, articulating with difficulty. As we well knew, this was a sign not of his admiration for the great poet, or of disorientation, but of a desperate effort to reassure himself that his powers of speech were not damaged and his faculties intact. After the brief recitation, he would turn imploring eyes to the doctors. He who had never wanted help, still less ever begged for it, now had to ask: "Doctor, I need three years to finish my work. See if there's anything you can do." The doctor assured Father that he was not in the least danger of losing his life. He would shake his head sadly and not reply. During one of these moments of recovery, he asked to be helped to an armchair near the bed. He did not want to die in the bed he had shared with Mother for thirty-two years. Following the advice of one of the doctors, we tried to stimulate his circulation by dipping his feet into a small basin of very hot water. I can still see Paola kneeling at his side, overcome with emotion, her eyes filled with tears, as she desperately rubbed a hot soaking towel on his legs and feet. And I can see Father's expression as, with infinite tenderness, he stroked her hair. The terrible attacks were followed by coma. We eased him onto the bed and, holding one another, watched his life slowly fade away. It was six o'clock on that summer morning, and the clear light of dawn lit up his face and ours, which were paler than the one lying on the pillow. I kissed his forehead, cold and still damp with the sweat of his final struggle, and thought with anguish of the kisses I had been unable to give him in my childhood. Over the large double bed where he was lying, while the tensed features were slowly relaxing into what is held to be the serenity of death, the bright eyes of a young woman, depicted life-size in a painting, smiled on him. Her nude adolescent body was stretched on the grass and surrounded by hovering *putti*. Just before their wedding, Mother had painted it at Father's wish, copying it from a seventeenth-century painting he loved. Since then it had always hung over their bed, an object of my infant admiration.

Gino sat in an armchair close to the bed, his head in his hands, and Nina, sobbing, fed her little ones, five-month-old twins whom she and her husband had brought with them in their dash down from the country. In tears, she breast-fed them; and they cried in their turn, little satisfied with the quantity of milk they were receiving from their exhausted mother. Our mother, who was lying on my bed in the next room, heard the long-dreaded news with speechless anguish and for several hours could not bring herself to approach Father's bedside.

More than fifty-five years have gone by since that night. And in that time I have suffered the inexpressible loss of three other people whom I adored since infancy: Aunt Anna, who died of cancer four years after my father; Mother; and Gino. Death showed itself to Mother and to Gino, however, in a quite different guise from that in which it had visited my father. Morris West, in the memorable first page of his *The Devil's Advocate,* writes:

> It was part of the decency of Death that he should come unheralded with face covered and hands concealed, at the hour when he was least expected. He should come slowly, softly, like his brother Sleep—or swiftly and violently like the consummation of the act of love, so that the moment of surrender would be a stillness and a satiety instead of a wrenching separation of spirit and flesh.

In spite of the immensity of the love that had bound me to Mother and Gino, the suffering caused by their loss was somehow softened by the sweetness with which they abandoned themselves to the arms of Brother Death. The pain that lacerated me with Father's departure, which had the same dramatic intensity that characterized his life, was due to the violence with which Death seized him and stole him away. Over all these years, my veneration for him has continued to grow, and even his death now appears to me as only commensurate to one who had lived so intensely and passionately.

CHAPTER 6

Medical Student: A Master's Apprentice

IT WAS in the autumn of 1930 that I entered for the first time the somber and stately amphitheater of the Institute of Anatomy of the Turin School of Medicine, situated on the tree-lined Corso Massimo d'Azeglio overlooking the Po—Italy's largest river, as they had proudly told us at school—whose cold green waters reflected the nearby glaciers whence it originated. The freshmen and sophomores who together attended the two-year course in normal human anatomy crowded the benches of the semicircular classroom, built according to canons handed down from the Renaissance when the first famous amphitheater for anatomy was built in Padua. In the middle of the amphitheater lay a cadaver. A young shivering freshman whispered to me that it was white because the blood had been drawn out with a syringe to make muscles and skeleton stand out more clearly. Before lessons started, the classroom resounded with ribald student cheers led by the bold sophomores and timidly echoed by the haggard freshmen. The two groups had in common only their fear and respect for the professor. Giuseppe Levi was legendary for the terrible rage he would burst into when the students continued to be rowdy and to sing after his entrance into the amphitheater. Pointing with the bamboo cane he used to indicate the object of the day's lesson on the blackboard and relevant parts of the cadaver, he would bellow: "Come straight to my room at the end of the lesson!" When the students presumed guilty came—penitent, with heads bent—to stand before the immense desk with its mountains of paper, he would ask them gruffly to explain their "abominable behavior," and then dismissed them immediately with a warm handshake and an "I beg your pardon, but you're a real pain in the neck."

48

Medical Student: A Master's Apprentice

That a lesson was approaching its end was indicated when the janitor, Conti, opened the amphitheater door. He was a figure no less legendary than Levi for his knowledge of anatomy, reported to be even greater than the professor's. Respectful but not obsequious toward Giuseppe Levi, Conti became advisor and consultant to all new students, who turned to him for secondhand textbooks and advice on how to prepare for exams. After the tremendously boring lesson was over (Levi's complete lack of speaking ability was aggravated by the fact that he detested macroscopic anatomy), the students, then about three hundred, streamed out en masse, the cheerful and vulgar goliardic songs would start up again, along with the "thrashing" of the freshmen by the sophomores and the older students in the more advanced courses who returned to the Institute of Anatomy just to participate in this ritual which lasted, in spite of Levi's anger, until Christmas. The girls—seven in my day and none of us very attractive—were split between first and second year, including my cousin and me. While we were certainly spared the thrashing, we could not avoid the less than gallant remarks about our aesthetic merits. One girl who lacked all charm was nicknamed "Greta Garbo in disguise" (I was relieved to discover it was not me) and brazenly pointed out every time she passed.

I felt inclined to spend time with the other freshmen and not so much with the girls, with the exception of my cousin Eugenia. Outstanding among the freshmen was a sixteen-year-old with a high forehead shaded by thick hair. His very dark eyes, inherited from his Calabrian mother, the almost hairless pink face, the short pants which revealed his muscular young legs, were all in contrast with his deep baritone voice. Neither he nor I were inclined to speak, much less to make friends, not least because since I had enrolled years late at the university and he very much in advance, the five years' difference in our ages was a further obstacle. The extremely brief conversations we had during the six years of university were a result of our both being, from our second year, Levi's interns and assigned an almost identical research topic which we both abhorred. Furthermore, in the third year, he transferred from the Institute of Anatomy to the Institute of Physiology, and our exchanges—from then until the day we took our degrees in the same July session—were limited to a few polite nods on his part, which I acknowledged still less overtly. Nevertheless, from the beginning of the first year, his physical appearance, his reserve—in sharp contrast to the loquacity of the others (for the most part, Piedmontese freshmen, facile jokers who talked in dialect and whose euphoria betrayed their joy at having overcome the difficult obstacle of high school), and the speed with which he solved the physics problems required in those days for medical students, all had awakened my curiosity in one so markedly unusual, coming as he did not from Piedmont but from Imperia in Liguria, where his father lived and worked as an engineer.

Not many months had passed before the rest of the students and the assistants became aware that this freshman possessed other qualities apart from the trivial ones just mentioned. The first-year examinations in biology, physics, and chemistry revealed the unchallenged supremacy he maintained throughout his student years. The successes he gained with indifferent ease, without the slightest hint of superiority, indicated an exceptional capacity for effortless assimilation both of the subject matter he preferred, like physics and physiology, and of those of no interest at all to him, like hygiene and certain clinical specialties still taught according to a tradition that regarded old nineteenth-century authorities as indisputable as Aristotle. In a later chapter, I shall describe how, many years later, this freshman— Renato Dulbecco by name—and I became close friends. He went on to win renown beyond the school and, in 1975, was awarded the Nobel Prize for Medicine.

I passed my first-year exams with honors and began the new term with commitment and enthusiasm. In our second year of medicine, Eugenia and I, Salvador Luria and Cornelio Fazio—who were one year ahead of us—and some other students following the same course of study, including Renato Dulbecco, all became interns at the Institute of Anatomy. We were attracted not by the subject matter, which held no interest for any of us, but by the extraordinary personality of the Master, Professor Giuseppe Levi, celebrated in Turin University for his reputation as a scientist, for the anti-Fascism he professed with supreme disdain for the most elementary rules of caution, and for the terrible but short-lived fits of rage I have mentioned already. We sophomores were assigned a bench in the room reserved to interns without any previous experience. United less by friendship than by a common interest in proving our ability, we would exchange precious histological recipes (not essentially different from culinary ones) which had been handed on to us with detached but benign superiority by the "old hands" who worked in a slightly larger room separated only by a door from our own. One of these older and more experienced students came every now and then to inspect the messes, the terrible histological preparations concocted by us novices. Other recipes were given to us, in exchange for a little money, by one of the three janitors, a certain Palmas, who was of subnormal stature, with a small sullen face and very black, extremely mobile eyes like little pinpoints. He had become quite expert in histology all on his own.

Half a century ago, at the time of my internship, histology was more of an art than a science, lacking the highly elaborate techniques available today. With the techniques then at our disposal, the best among us would obtain satisfactory results in the preparation of tissues from laboratory animals, from biopsy or autopsy. These preparations consisted of slices a few microns thick (a micron is one thousandth of a millimeter), fixed, stained, carefully arranged on glass slides, and examined through an optical microscope. This microscope was improved but not essentially different in construction from the first such microscope used around 1660 by the great Bolognese biologist Marcello Malpighi for studying the structure of vegetable and animal tissues.

The difference between a successful or an unsuccessful histological preparation depended, and still depends, not so much on the meticulousness with which one follows the rules set out in the manuals of histology—which we jealously guarded, along with the recipes obtained from the old hands and from Palmas, in the locked drawer of our bench so that they wouldn't be meddled with or borrowed and not returned by the other interns—but on whether one possesses "green fingers." From the start, it was clear that I had no such gift. I became aware of this failing when the best of the interns and Levi's favorite pupil, Rodolfo Amprino—younger than I as well as two years ahead, and "most gifted student," as the Master would proudly declare—used to frown and shake his head whenever he inspected my slides under the microscope. I never imagined then that he was to become my closest friend during the difficult years, nor that I would be able to rehabilitate myself both in his eyes and in those of the Master thanks to a certain ability I showed in applying the so-called silver-impregnation technique on nervous tissues. This technique was introduced toward the end of the last century by the great histologist and founder of the science of neurology, the Spaniard Santiago Ramón y Cajal (see pp. 126–27).

During the first year of our internship, Levi had assigned to many intern-pupils—including my cousin Eugenia, Cornelio Fazio, Renato Dulbecco, and me—the task of investigating whether the number of nervous cells recognizable from their topographical characteristics was constant, as had been ascertained by counting those of entire nervous systems in lesser invertebrates; or whether, on the contrary, it varied in small mammals, such as the mouse, which lend themselves easily to this kind of inquiry. Levi also wanted to determine whether the number of cells in the sensory ganglia (small ovoidal formations located to the left and the right of the spinal vertebrae), which range in the mouse between ten and twenty thousand depending on the spread of innervation, was the same in mice from the same brood or differed in animals from different broods. Eugenia and I had been assigned this latter problem, which was not only tedious but also left us doubtful about the validity of the counts made with such rudimentary techniques.

One spring afternoon of that year, 1932, while Eugenia and I were busy counting the nerve cells of spinal ganglia, Giuseppe Levi barged into the room with Tullio Terni, then professor of anatomy in Padua and one of Levi's first and most esteemed pupils. Tullio Terni differed in every respect from the Master. Levi was tall, and his robust constitution was strengthened by his passion for mountain climbing, a sport he pursued summer and winter as tenaciously as he pursued problems in biology. His thick red hair, which had inspired his friends to give him the affectionate nickname "Levipom" (*pom* being a diminutive of *pomodoro,* "tomato"), the bushy eyebrows which almost entirely hid his greenish-brown eyes, his way of walking with head bent slightly forward, and his complete indifference to dress made one think of a Russian—anything from moujik to philosopher, careless of the world and absorbed in his own thoughts, or a writer of the Tolstoy sort, landed in our midst by mistake. Terni, on the other hand, was of average height, delicate in constitution, with handsome, regular features and sharp eyes. He was always well dressed, and the monocle he lodged under his eyebrow accentuated the nineteenth-century elegance of his delicate and slender figure. In cultural interests and political opinions, the two men differed even more than in physical appearance and bearing. Levi not only did not hide his scorn for the

Medical Student: A Master's Apprentice

Fascist regime, its leader Mussolini (the Duce), and the buffoons in the hierarchy but enjoyed proclaiming his ideas to acquaintances, colleagues, or assistants whom he met on public transport (which happened frequently since he didn't use a car). Happy at the encounter, Levi would loudly proclaim his opinion of the latest Fascist idiocy to appear in the newspapers exalting the Duce's genius, while his unfortunate victim would nod in mute embarrassment, intent on thinking of an excuse to get off at the next stop. It is significant that, although the great majority of citizens wore Fascist badges in their lapels, there was not among them an informer—that shoddy species of humanity which flourished a few years later in Nazi Germany; nor were Levi or any of his involuntary interlocutors ever forced to drink castor oil, the remedy the Fascists found to cure citizens suffering from subversive tendencies. It, thus, appears that enrollment in the Fascist party was compulsory and that, in their hearts, the accidental listeners undoubtedly shared Levi's opinions. This fundamental difference between the Fascists and the Nazis as a whole was later demonstrated by the jubilation that greeted the unexpected radio announcement in July 1943 that His Majesty, the King Emperor Vittorio Emanuele III, had accepted the resignation of the Prime Minister, His Excellency the Cavalier Benito Mussolini. Terni, on the other hand, did not dislike the regime, an attitude that infuriated the Master and seemed to me to stem more from a desire to goad and enrage old Levi than from real adherence to the Fascist ideology.

In terms of cultural interests, the arts, music, literature, and philosophy, Levi and Terni's differences were equally great. The former had no appreciation for either the figurative arts or music; and in literary matters, I think that—though I never discussed the point with him—his knowledge and interests were limited to classic works in Italian or German, a language he knew as well as Italian. Terni's deep and extensive knowledge of and passion for art and literature were, on the other hand, manifest even in the course of brief conversations which were sparkling and full of Proustian undertones. Despite his excellent scientific work, which Levi greatly appreciated, Terni's brilliant and versatile mind might well have been more at home in the figurative arts or the humanities rather than in the science he chose. The differences in personality, character, and opinion between these two men did not, however, prevent them from feeling great respect for one another. Hence, Terni's frequent visits to Turin.

It was during one of these visits that our first meeting took place. After briefly introducing us, Levi informed Terni of the project he had assigned and of the results so far obtained. While listening to what the Master was saying, Terni was somewhat distracted by Eugenia, whom he watched closely through his monocle. Her mischievous, smiling face was attracting him far more than my serious, almost sullen expression. Noticing this, Levi pulled him away by the sleeve, saying that he wanted to show him Rodolfo Amprino's beautiful preparations in an adjacent room. Terni, however, asked Levi to let him stay with us a little longer to examine our histological preparations under the microscope. As soon as the Master had left, Terni looked at the preparations for a few minutes and then fixed us with a witty smile. "After you've finished this research," he said in a low voice so that Levi in the next room would not hear, "you could start another project, no less interesting. With the help of a ladder and the right instruments, you could count the number of leaves on the branches of the two plane trees outside the window, and then write up the results with your count of the number of leaves on the tree on the right and the tree on the left. It's possible that you won't manage to decide whether any eventual differences are to be ascribed to the varying exposure to sunlight or to other causes, but that is the fate of ninety-nine percent of the research we do, including the work you're now engaged in." He remained silent for a moment, wondering how he might soften the effect of what he had just said. "I admire the Master very much," he resumed, "because he is a great man and a great scientist, but I would advise you in the future not to accept projects like this one."

We and the other interns had been wondering for some time why Levi wanted these extremely tiresome counts. Certain interns, with a sincerity that shocked me, had admitted that they made up their figures, not only saving themselves toil they deemed entirely useless, but also making sure the professor would like the results. In the case of Eugenia and myself, the counts, which we actually performed, confirmed Levi's forecasts, and he quoted them in his great treatise on histology, though we still had doubts about their reliability. More than half a century later, as I think back on the problem that interested Levi, left Terni entirely skeptical, and made us unhappy— basically because of the tedium and the effort involved in so much counting—it seems to me that we all were right and wrong at the same time. The problem was in itself much less futile than Terni thought. Its aim, in fact, was to determine whether the number of cells in specific and clearly identifiable nervous groupings is rigidly fixed and not subject to fluctuations as a result of environmental factors. The same problem has been tackled by illustrious biologists in recent years and is still far from being solved. Half a century ago, however, the techniques available to us were too primitive to resolve it, and we knew from experience how chancy our figures were. In proposing it to us, Levi was much ahead of his time, but the fact that it was impossible to solve justified both Terni's skepticism and our reluctance in carrying it out. Levi had also made a psychological mistake in proposing it to novices such as ourselves. It is not hard to overcome the tedium and effort required to count tens of thousands of cells when one has clearly understood and grasped the significance of the problem, but it is much more difficult when one is entirely in the dark about it as we were.

I saw Terni again briefly in the following years when he came to visit the Master, but had the chance to spend more time with him in 1936, four years after our first meeting, when Eugenia and I attended an international anatomy conference in Copenhagen with him and Levi. For us it was the prize trip for having gained top honors in our degrees in medicine. During the brief pauses at lunchtime or in the evening, when the day was over, the four of us would walk in the park or along the avenues surrounding the conference center. I remember one such occasion when Terni, affectionately arm in arm with Eugenia, was strolling a few yards in front of Levi and myself, who kept a respectful distance apart. Terni was singing: "Blossoming gardenia, there are many deer in the world, but only one gazelle, Eugenia." He lengthened out the *g,* as he gazed at her through his monocle. Levi, who appeared to be silently absorbed in his thoughts, suddenly put his big hand on Terni's shoulder and gruffly asked, "What are you saying?" "I was pointing out to Eugenia," Terni answered, straight-faced, "those yellow children playing ball." He meant not the skin color but the almost albino blond hair of some small Danish children nearby who had nothing to do with either deer or gazelles. Levi scowled and grumbled, "Idiocy!"

Upon our return, Levi, having first made sure that Terni and Eugenia had not yet come into the conference hall, sat down next to me and expressed his concern in what he thought was a low voice: "Tell me, don't you get the impression that that ass Terni, is courting your cousin? She ought to be given a word of warning, because Terni is a great womanizer." I should say that the term *ass,* which we think of as being very insulting, was not so for Levi, any more than were the other terms, *imbecile* and *pain in the neck,* which he used without stint. While greatly respecting Terni's intelligence, Levi disapproved the interest he showed in young, beautiful, or at least attractive women, like Eugenia. I assured Levi that there was no such danger, that Eugenia had always been cheerful and spirited by nature but was also extremely responsible, and that her responding to Terni's gallantries did not imply an infatuation on either side. Indeed, while a keen observer, Levi lacked psychological penetration.

When I saw Terni again at the end of the war, he was in a state of deep depression. Such crises, frequent with him, had become more severe in recent years. He was in this state of mind on 4 January 1946 when he received a letter of expulsion from the Accademia Nazionale dei Lincei. The National Academy—having been suppressed in 1939 by Mussolini, and then reconstituted in April 1945—had sent letters of expulsion to no less than forty of its members because of political conduct during the Fascist regime. Among them were some prominent exponents of the Fascist party, such as Cesare Maria De Vecchi di Val Cismon, Luigi Federzoni, Alberto De Stefani, Giuseppe Bottai, Franceso Severi, Sabato Visco, and Gioacchino Volpi. The rest were professors who had adhered to fascism in the same way as had the great majority of Italian citizens. The very fact that such a notable percentage of academy members (nearly a third of the total number) were subject to this punishment shows that the decision, the outcome of lengthy discussions in the autumn of 1945, made no distinction between people who had been guilty of terrible misdeeds and others who, in a less inflamed atmosphere, would have at most deserved a mild note of censure. One of the commission's six members was Giuseppe Levi, its acting secretary. When he told me that Terni would be among the expelled, I had pointed out the danger that such a sanction—which I believed to be entirely unjustified, since Terni had never occupied a political post and had done no more than sympathize with the regime—would aggravate his depressive state. Levi instead felt it to be his duty, precisely because Terni was a personal friend, to proceed against him in the same manner as all the others. As far as I know, the other thirty-nine expelled took the matter philosophically. All of them, I believe, were reinstated two years later, and many of them wound up in prominent positions. One became dean of the University of Rome and enjoyed great prestige; another enjoyed no less prestige as a mathematician in the same university. Only Terni, who was already depressed, fell into a state of despair which caused him to take his life on 25 April 1946. He used a vial of cyanide which he kept on hand, like the ones he had given to his family at the time of the Nazi invasion, in the reasoned belief that if they were ever captured by the SS it would be the wisest and quickest solution. The news of his death reached us at the institute the following morning. It was a terrible blow for Levi, who realized too late how he had underestimated Terni's enormous sensitivity and vulnerability. From then on, Levi kept on his desk a large photograph of Terni, who seemed to regard the Master with a witty and deeply sad smile.

To return to my second year of medical school, when, with neither glory nor disgrace, we had finished counting the nerve cells of the spinal sensory ganglia in mice from the same and different broods, Levi called Eugenia and me into his office to assign us another research topic. Eugenia's lot was to study the development of the tonofibrils: that is, the fibrillary structures in the epithelial sheet covering the hoofs of calf fetuses. She came out of the office satisfied with her task; though somewhat dull, it seemed to present no difficulties, as proved to be the case. Then came my turn. Levi greeted me with a brisk nod of the head and set himself to look for a problem to entrust me with. He skimmed through his notebook entitled "The Hammer's Sparks," a title that revealed his unconfessed and secret passion for the poet Gabriele D'Annunzio who, through his exploits and derring-do following World War I and his poems celebrating Italy's past glories, had become the favorite bard of the Duce and of the regime. The Master's large finger stopped at a topic that had been underlined, showing he had already given it more than passing consideration. "Here's a topic suited to you," he said. "You will study how, and by means of what processes, the convolutions of the brain of human fetuses are formed." Despite my total lack of experience, I realized that it was a task destined to total failure. Outside, the other interns were waiting to know what topic I had been assigned. "He managed to unload it on you," one of them commented in a pitying tone. "He tried it with me, but didn't succeed." Though I suspected he was merely boasting—no one ever dared question the Master—this comment from an older student confirmed my pessimistic forecast. With furious determination, I set myself to tackle a problem that proved from the start far more difficult even than I had feared.

The first stumbling block was how to obtain the necessary material. How to find, in the early 1930s, human fetuses from an early stage of gestation? There were no official abortions in those days, and I could not resort to the enormous number performed clandestinely by doctors or, as in most cases, by unscrupulous midwives ignorant of the most elementary rules of sterilization. Apart from these clandestine abortions, the only other source was the very poorly preserved material deriving from spontaneous ones, or the ones occasionally provoked, when a mother had serious problems, in the maternity ward of the city's Ospedale Maggiore. The caretaker, enticed by the handful of lire promised, called me a few days after I had turned to him for help: "I've got what you need," he said. I rushed over and was unable to conceal my disappointment when he brought me what was, to all appearances, the corpse of a newborn child. "But this is a baby, not a fetus." "Take it or leave it." He wrapped it up in a bundle of newspapers and handed it to me. Carrying the heavy parcel, I got on the first streetcar going straight to the Anatomy Institute. During the ride I was suddenly horrified to notice that a little foot was sticking out of the bundle of newspaper, and trembled to realize how suspicious I, just a little over twenty, and my burden would look to any passenger who had noticed.

I hastily got off at the next stop and continued to the institute on foot. I was met by the janitor Conti, the autodidact, counselor to the students, sharer in our anxieties. He had shaken his head on hearing the topic assigned to me. He examined with expert eye my fetus-baby. "But this one's fully developed," he commented, "and over three days dead. It won't get you anywhere." He helped me to dissect the brain, commenting on my recklessness in traveling by public transport with that baby in my arms: "You should thank your lucky stars you're not in jail." The brain fell apart under his instruments. The already formed convolutions, which would never have functioned, fell apart under the bistoury before I could subject the slightest fragment to histological examination.

In the following months, though I had learned not to use public transport when carrying material from the maternity ward to the institute, other failures followed upon the first. The professor called

58

me in to examine my preparations which were supposed to elucidate "how the convolutions of the brain are formed"; he called them "real trash" and came to the conclusion that I was decidedly not cut out for research. I was automatically blacklisted as one of the students whom he defined as a "pain in the neck," a definition all the more distressing to me since, while my stock was plummeting, Eugenia's was soaring. Calf fetuses were obtainable in large quantities at the slaughterhouse, and the tonofribrils could be easily identified by means of common histological techniques. While her research was proceeding at full steam, my lack of success convinced Levi that I was completely inept.

I was saved from this critical situation by having to be admitted to the clinic for an emergency operation. For a month I was forced to abandon those contemptible convolutions. It was typical of the Master that he would rush to the bedside of his assistants and students when they fell ill. When I returned to the institute, he allowed me to abandon the project and to start on another topic. This one, unlike the two preceding ones, gratified me immensely and marked the beginning of a Master-disciple relationship, characterized by ever-increasing affection and reciprocal esteem which lasted until his death thirty-one years later.

This research gave me my first glimpse not of the much-feared professor but of a master who had a real passion for his work, a critical sense far superior to that of the majority of biologists of the day (with the exception of Terni and a few others). For the previous six years—since 1928, that is—Levi had devoted himself with increasing enthusiasm to the study of nerve cells grown in vitro. This research had begun in the United States during the first decade of the century with Ross Granville Harrison's investigation of the growth of nerve fibers from amphibian spinal-cord fragments in a semi-solid medium, and had been extended to other tissues thanks essentially to Alexis Carrel, author of the famous book *L'Homme, cet Inconnu*, and to a few of his disciples. Showing considerable intuition, Levi had immediately realized the enormous possibilities opened up by the application of this new technique (which most considered a mere curiosity) for the analysis of the proliferation and differentiation of cells from different stock in an environment controllable by the experimenter, as was possible with in-vitro cultures. The numerous articles by Levi and his co-worker Hertha Meyer on sensory nerve

59

cells grown in different experimental conditions are major contributions in this then just beginning area of research.

Levi entrusted Eugenia and me with a topic that became that of our doctoral theses as well: the formation of the collagene reticular tissue by the connective, muscular, and epithelial tissues. For the first time, I became passionate about research, and the enthusiasm with which we gave ourselves to this new project did not go unrewarded. We succeeded in demonstrating for the first time that the formation of reticular fibers, which can be revealed by a particular argentic staining, is not a property of connective tissue alone but of muscular and epithelial tissues as well. Up till then it had, in fact, been believed that collagene reticular fibers, namely, the fibers that make up the weft that supports the complex of tissues of which they are a part, were produced by special cells present in great numbers in bone, cartilage, and subcutaneous tissues alone. Though I did not have further occasion to continue that research, the experience I gained studying tissues in vitro, and especially nervous tissue, was to prompt me later to resort to the same technique in uncovering the then completely mysterious nature of the Nerve Growth Factor. I shall return, in chapter 17, to the subject of that research, which marked an extremely important milestone in the discovery of the Factor.

Before turning from the influence that Levi had on me, I should like to recount two episodes from my early apprenticeship with him. The first reveals the Master's comical nature; the second, his intransigence. Once, during a laboratory session, Levi wanted the students to examine cells that had been freshly peeled from the surface of the oral cavity. Saliva was obviously the readiest and most economical source for such a study. Holding a carefully cleaned microscope slide, one of the assistants approached him and said in a deferential tone: "Spit, Professor." Levi burst out laughing and, as he would often do, gave a heavy slap on the back with his large, hairy hand: "You spit. A professor's saliva isn't worth more than that of a student or assistant." Though quite embarrassed by the general laughter, the assistant did as he was told. The second episode occurred while a student was defending his doctoral thesis and I was awaiting my turn to do so. The candidate before me had compiled his thesis with the institute pathologist, who was equally renowned for his ability in the operating room and lack of general culture. The thesis was on mitogenetic rays, which had been discovered by the Russian biologist Alexander Gurvich who maintained—if I remember rightly—that such rays emanate from certain plants and stimulate the proliferation of animal or vegetable cells exposed to them. Suddenly Levi erupted and, unrestricted by the academic gown then customary on such occasions, slammed his fist on the table, declaring that he would not stand for such nonsense and would not accept the thesis. The pathologist stood up and left the room, livid with rage. Subsequently he claimed that he might have stood for such remarks in private, but could not accept that this Jew should express himself in such a way in public. When Levi heard about it, he shrugged and said, "I don't give a damn what he thinks, and I'm very happy I said what I said."

CHAPTER 7

Fellow Students: Friends for Life

I DON'T KNOW whether the little inclination I felt to make friends at the university with the many young men and the very rare women in the same courses as I is to be explained by the fact that the longed-for study of medicine seemed incompatible with the extracurricular activities preferred by the great majority of students in the 1930s. I recently chanced to meet Gigi Magri, one of the few fellow students with whom I had been acquainted—though always from the required distance. "I hope you won't take offense," said Gigi, "but you are pleasanter now, despite your age, than when you were young. Then you were just impossible, a kind of squid ready to squirt ink at anybody who came near you." This joking comment of his, which I'm not sure I should have taken as a compliment, conjured up for me the image of the young woman I had been, with her grim face and her clothes almost nunlike in their severity. Though it had not been in my intention to squirt ink, this evidently had been the impression I gave my fellow students. A few, nevertheless, tried to overcome the barrier I had built around myself.

The first of these was a young second-year student, small in stature, very fleshy. Immersed as I was in my own thoughts, I did not usually take notice of the people getting on and off the bus I rode every morning from home to the Anatomy Institute building. Nevertheless, after the third apparently casual encounter with the young man, whose name I didn't even know, I became suspicious. The fourth time I got off at the stop preceding the one he had boarded on previous days and walked to that stop. From a distance I saw that, instead of boarding, he was anxiously examining the passengers from

behind a tree. Not seeing me on this occasion, he signaled the driver
to go on without him. From then on, I changed my route, but S. did
not give up.

During one of Levi's lectures, S. slipped me a note and whispered
that he wanted my opinion of what he had written. "You," he said,
"are a young lady, but I hope you won't be distressed by the strong
and manly expressions. In poetry, as the great poets teach us (and he
mentioned two of the best known), we must use them if necessary."
On the piece of paper, S. had written what he took to be a sonnet in
which he laid bare his mind "in a manly way." So much did it amuse
me, that I still remember it.

O sloven heart
why do you weep in rage?
Tormented, why cry out
against nature
who with horrid grin
mocks at you and laughs?

He went on to catalogue his sufferings over a young woman he
adored who didn't requite his love, and finished with a scream, curs-
ing fate for mocking him. I returned his sonnet telling him how sorry
I was for his pain, but that I had no help to give. Timidly he asked
me another favor. His parents had opened a pastry shop in one of the
busiest streets in the city. Could I see to it that we used their service
for the "at homes" that certainly must often be held at my house? I
regretfully told him that I could be of no help in that either. Contrary
to what he thought, and this was really the case, no receptions were
given in our house. Sighing, S. gave up his amorous and gastronomic
approaches, and I was able again to catch the bus that took me more
directly to the institute.

There followed an experience that was to leave a much deeper
mark. The person in question this time was a first-year student, extremely tall, with a penetrating gaze, a musketeer mustache and a
curious way of walking, head down and always whistling a Beethoven symphony or an aria of Schubert's or Mozart's.

He was the intimate friend of a cousin of mine, who was also a
fellow student, and, thanks to him, overcame the barrier I had set up.
From the very first day of lessons, Guido had shown interest in me,
and I could not be as offhand with him as I had been with S. I

accepted his invitations to go on long evening strolls in the Valentino Park, on the condition that we talked only of cultural and musical subjects. I beat him in knowledge of literature and the arts; he was a long way ahead of me in music. When I listen, more than half a century after those strolls of ours, to Schubert's C Major Quintet or Beethoven's Concerto in D Major for violin and orchestra, I relive those hours spent in the park or on outings when I listened to his fine bass whistling of Mozart and Brahms and tried not to fall off the pillion of the motorcycle he drove skillfully but fast up the steep stony tracks of the valley. My friendship with him was to last through the years following graduation and the war. During the Nazi occupation, he played a heroic part in the underground struggle, miraculously escaping capture.

The other friendships of my university years were formed in the Institute of Anatomy, in the rooms assigned to us as members of the Department, and in the big library that was the pride of the institute and particularly of the professor who passed long hours there.

All members of the Department were allowed entry to the library, but it was not easy for any of us to consult the books. Where nowadays periodicals pour into scientific libraries, they were scarce in those days; and I do not remember what the arrangement was for recent issues. The older bound copies and works by authors going back to the mid-nineteenth century took up enormous glass-fronted bookcases of solid wood which reached to the ceiling. To consult the books—mostly in German, the language biologists preferred to write in up until the 1930s—one had to climb a wobbly ladder. Since it wasn't permitted to take books out of the library, we had to consult them on one of the large tables in the middle of the room and replace them in the book cases immediately afterward. The temperature on cold winter days was kept down to 12° C. in order to discourage "fans of science," as Levi called the most zealous and diligent students, and even more to keep out the "lazybones" and "pains in the neck," as he labeled those he thought had no scientific aptitude or interest but used the library as a place to meet and gossip. Anyone leaving an overcoat or personal belongings on the tables was in for trouble. I recall one tremendous outburst from Levi when he bumped into some poor chap who had used one of the tables for just that. In a thunderous voice, Levi reminded him that the library was not a bar meant for uproarious behavior. Propelled by the professor's contemptuous glare, hat, overcoat, and briefcase fled hastily, along with their unfortunate owner, from the holy place they had desecrated. I remembered this episode many years later when I first entered the library of the biology department at Washington University. It was full of students in shirtsleeves. Many of them were lying back in the comfortable chairs, their shoeless feet propped on the table tops, chewing gum while they read periodicals or, tired of reading, slumbered deeply, pillowed on reviews and notebooks.

It was in the laboratory or the library that meetings took place between myself, Luria, Dulbecco, Fazio, and Amprino. The latter, who—despite his exceptional youth—had been Levi's intern at seventeen and was twenty at the time of our meeting, was the professor's favorite and remained so until the day of his death. Amprino (whom I addressed by his surname though he, in his brusque way, called me Signorina Levi) somewhat awed me with his well-known skill in histology and knowledge of the subject, light years ahead of mine.

My friendship with Amprino, however, as well as with Dulbecco, did not really flourish until right after the war, as I shall describe in chapter 13. The brief conversations I then had with Dulbecco and Fazio were limited to exchanging information about the results of our cell counts in the spinal ganglia of different animal species. Dulbecco, bored and disappointed with the task set him, announced his results without interest in a low voice, determined as he was to leave his internship as soon as possible—as he did, to enter physiology, the following year. Fazio, more cheerful and talkative, would go on at length about the results of his cell counts in the ganglia of amphibians of the species *Pelobatus fuscus* and *Rana esculenta*. Luria, who had been assigned a different project, looked at us with profound commiseration. There was, moreover, a kind of competition between him and Cornelio Fazio caused by the professor's open preference for the latter. "I like Fazio," the professor told me two years later, when greater trust had grown up between us, "for his intelligence, but even more for his human gifts." He recognized Salvador Luria's (Salva, as I came to call him) intelligence but defined him as "petulant." Luria, who was aware of the professor's view of Fazio and himself, would let off steam with me. "It isn't his so-called human gifts that gets him in with the professor, but the fact that his father is a socialist and has a name as an anti-Fascist. And what does he know about me and

65

mine? Tomorrow I'll see to it that he accidentally finds a little note on my desk, saying 'Kisses from your darling Stalin,' and then you'll see how he'll change his mind about me."

In the years following their graduation, a year before my own, I rarely heard about Fazio or Luria. Fazio participated courageously in the war as a partisan. When the war was over, he took up his work as a scientist again and rapidly made good the lost years, dedicating himself to studying circulatory disturbances in the central nervous system. I shall return later on to Luria and to his remarkable scientific career.

In our first two years at the university, our time was mostly taken up with lectures in the mornings and with afternoon examinations in macroscopic anatomy, which took place in the large rectangular hall next to the anatomy amphitheater equipped for dissection. Almost every one of the ten tables held a cadaver or part of one which, despite the generous use of disinfectant, gave off the unmistakable odor of decomposing tissue. In charge of dissecting was Professor Li Causi, a Sicilian of noble extraction, as one then used to say. So noble was he, according to people in the know, that he had never allowed his wife, much younger than he and of less noble origins, to sit down to dinner with him. She served him and then waited respectfully behind the door; only when he had finished was she authorized to eat. Li Causi's face was stern, his gown immaculate, and he had a passion, inexplicable to us, for dissecting cadavers. He never wore gloves on the slim aristocratic hands that handled the instruments with the skill and grace of a virtuoso violinist, and his eyes sparkled with pride and joy when he showed us a small empty triangle between two dorsal muscles in the lumbar region. Modestly he would tell us that this was the triangle he had discovered. In fact, in textbooks of normal human macroscopic anatomy, it bore his name. "That imbecile Li Causi's famous triangle" was how Levi disrespectfully referred to it, breaking into a bellowing laugh, whenever it was mentioned. In reality, Li Causi was by no means an imbecile but a refined artist-craftsman, passionate about his art like all craftsmen, and extremely capable in his line of work. We, in our turn, with our clumsy, unskilled hands and ill-disguised revulsion, did our best, with little success, to perform the same dissection on the cadavers lying on the nearby tables. The majority of students didn't wear gloves either, which were thought to indicate a womanish fear, though they took great care not to cut themselves; I don't remember any accident occurring. I, the privileged bearer of two X chromosomes, never engaged in the detested task without a fine big pair of rubber gloves which reached up to my elbows.

It was in the course of a dissection exercise in the anatomy amphitheater that I met Germano R. Raising my eyes from a corpse, I found his eyes on me: timid blue eyes which looked at me while, without using his hands which had touched the corpse, he kept tossing back his blond hair which constantly fell forward again. He introduced himself in an embarrassed fashion, and from his accent I recognized his Lombardy–Piedmont origins. He came from Volladossola, in the province of Novara, a little mountain village near the Swiss border where his father was district doctor. His elder brother Luigi, whom he pointed out, was busy with a dissection some tables away. Luigi, a year ahead of us and the object of Germano's admiration, was watching with amusement his younger brother's first efforts at approaching a woman. Germano that day set about a timid and discreet courtship which lasted through all the years of university. Never on his part, and even less on mine, was there any hint even of the possibility of marriage. He believed, I think, that such an eventuality was made impossible by the great difference in our origins. His mother, as he told me with a mixture of embarrassment and pride, could hardly write her own name and had married the village doctor because he had been attracted not by her healthy peasant beauty but by her timidity and submissiveness.

We graduated together in the summer session of 1936; and when Germano learned of my choice, he too decided to specialize in neurology and psychiatry. Our relationship, however, remained one of courtesy and detachment. In 1938, when the anti-Semitic campaign was taking on an increasingly threatening tone in the newspapers and in the speeches of Fascist officials, Germano worked up his courage and boldly declared his love for me and his hope of sharing life together. From being an enrolled member of the Fascist party, as were the vast majority of young men in those years, he became a ferocious anti-Fascist, more indignant than we at the vicious campaign against the Jews. He began coming to our house, drawn by a strong affection for Mother, Gino, and Paola (our married sister no longer lived with us). The affection was mutual, and for the first time Germano spoke openly about the possibility of marriage—a proposal received with

67

emotion by Mother and Gino, who were amazed at his behavior and the courage with which he defied the certain consequences for his career. The decree of 17 November 1938 prohibiting marriage between Aryan and Jewish citizens, however, obviated the possibility which, though much desired by him, did not correspond with my wishes. I esteemed Germano highly, in fact, and was very fond of him, but our differences in temperament and in cultural interests seemed to exclude the possibility of marriage.

At about this time, Germano's health began to deteriorate, with intermittent fevers accompanied by a state of profound lassitude, and by the following year was so bad that a doctor was called in. The doctor diagnosed tuberculosis of a miliary kind. At that time, because of the racial laws (see p. 82), I had been dismissed both from my academic position at the Institute of Anatomy and from the Neurology Clinic, and deprived of the right to practice medicine. I decided to accept the offer of Professor Laruelle, director of a neurology institute in Brussels, to continue my research in his laboratory. In March 1939, I left Italy for Belgium. Germano, in a feverish state, accompanied me with Paola and Gino to the border at Bardonecchia. In taking leave of me, his eyes full of tears and hardly struggling to hold them back, he said, "Rita, this is the last time we'll see each other. I haven't much time to live." His pessimism seemed exaggerated to me. He was twenty-eight, still young, and of tough mountain temper. He would get over the crisis, and we should soon see each other again. Waving from the platform as the train left, he shook his head.

In Brussels, every day from then on, I received a letter from him. He kept me informed about his gradually worsening health caused by the addition of a serious infection of the kidneys, tubercular in nature, which gave him a lot of pain. When I heard that a sudden worsening of his condition had caused him to have himself taken to Villadossola so that he could die in his parents' house, I left Brussels at once. His last letter, handed to me when I reached Villadossola, remains one of the most dear and painful documents I possess—yellowed by time and, as he wrote, by tears. I shall quote a brief passage:

Dear Gino who gave me the machine which measures time (a wrist-watch) will allow me, in case I make a long journey to where the hours don't count, to give it to Mother or to Rita. What I have received from You [always capitalized, such was his love and respect for my dear ones] is so dear to

68

me that I guard it with fierce jealousy and true religious feeling, and for me, who puts no value on anything, it has so great a value that only by you can it be repossessed.

I got to Villadossola with Gino on the evening of 23 July. Although I hoped to see Germano immediately, his father preferred to postpone the meeting till the following morning in order not to upset him. When we arrived, he was still conscious but was no longer so when I was able to enter his room early in the morning. He squeezed my hand tight without recognizing me and murmured indistinctly something about Galeazzo Ciano, Mussolini's son-in-law. I could not understand the one or two other phrases he muttered. The thought of the persecution our family was undergoing distressed him to his last breath. I was present when he died that same evening, free at last of suffering. Two days later, in the church he had so often described to me, I knelt with his family in the ritual with which the living say goodbye to those going on the long journey to an unknown destination where a machine for counting time is no longer necessary.

PART II

The Difficult Years

CHAPTER 8

Premonitions of Trouble

ON 11 JUNE 1924, Italians were shocked to read in their newspapers that the Socialist deputy Giacomo Matteotti had mysteriously disappeared while walking to Parliament, where he was to have made an important speech criticizing the government. The newspapers suggested that Matteotti had fled the country, but the testimony of two implacable and daring ten-year-old eyewitnesses and the license number of a mysterious motorcar, which had been patrolling the area where he disappeared, cast grave doubts upon this convenient explanation. Renato Barsotti, nicknamed Neroncino ("little Nero") for his pugnacity, and his friend of the same age, Amilcare Mascagni, were playing at war on the banks of the Tiber river when they were advised in a bullying manner to move on by a powerfully built man who had gotten out of a car with four henchmen. A backhanded slap in the face persuaded Neroncino not to ask questions or protest. A few moments later, they saw the five pounce upon a man who had emerged from a nearby street with a briefcase under his arm. Though he fought desperately to free himself, they overcame him and forced him into the car, which then took off at full speed in the direction of Ponte Milvio. Pasquale Villarini, the watchful doorman of a nearby apartment building, had taken down the license number of the car, which had roused his suspicions because of the way it was patrolling the area with five men on board. Villarini, Neroncino, and his pal thought it their duty to report what they had seen to the police. Their statements, accepted with the distrust and annoyance described in the memorable book *Quei Vent'Anni (Those Twenty Years)* by Matteo Matteotti, the victim's son, nonetheless permitted the tracing of the car and its owner, the Fascist gang-leader Amerigo Dumini. Neroncino, outraged not least for having been slapped by Dumini, returned to the site of the kidnaping and painted there a large red cross. This became, in subsequent days, a place of pilgrimage where friends and admirers of the young murdered deputy laid flowers. Today, instead of the childish cross, a bronze spear-shaped structure points to the sky. Every day when going to the lab, which is only about a hundred yards from the spot, I think with regret of that rough vermilion cross, and its replacement by a small monument similar to the thousands of others in all public squares of this and other countries, which remind posterity of the glorious and inglorious facts of their history. The cross painted by Neroncino should have remained to remind us, the survivors, of Italy's first act of surrender to brute force, as the first tangible sign of the corrosive process that would lead to its downfall.

A few weeks later, Giacomo Matteotti's mother and his wife—Isabella and Velia Matteotti, respectively—went to the Palazzo Venezia to implore Mussolini to order a search for his body. The subsequent discovery was macabre, the poor body having been reduced by predatory animals to a contorted skeleton with just a few scraps of flesh still attached. The extreme brutality of Matteotti's murder, reported at a time when there was still freedom of the press, and the farce of a trial, which Mussolini tried to cover up by having it held in a small provincial town instead of Rome (Dumini received a mere five years and was then released almost immediately thanks to a royal amnesty), marked the most critical moment for the Fascist party since the Duce had taken power in October 1922. Mussolini's "madman's eyes"—which Isabella Matteotti described so vividly to her young grandson, the eyes that exerted a hypnotic power on the masses crowding the public squares—betrayed on the day of the interview the fear of someone caught red-handed: it was all too evident that Dumini had been acting under orders. A letter of this hired killer, reproduced in Matteo Matteotti's book, clearly reveals the relationship between the executor of the crime and the man behind it. The letter, written in the Roman prison of Regina Coeli on 8 April 1926, two weeks after the farcical sentence, carries, as Matteotti points out, a curious letterhead: "Daring Fascist Group—Regina Coeli Section." "Strange prisons," Matteotti comments, "those existing in Italy at the time, where persons condemned for homicide might consider their cells the offices of a well-regarded institution with its personalized stationery." The handwritten letter, revealing the complicity of the relationship, is worth reproducing:

> Duce! In this hour so painful for us, but so triumphal for Thee, allow me to send You, on behalf of my companions also, all of my deeply feeling and devoted soul. Though our sentence has left us indifferent, we felt its enormous weight yesterday when it prevented us from seeing You and hearing You and uniting—in freedom—our congratulatory cheer to that of all the Italian people. Duce! All our faithfulness to Thee, all our trembling soul; all our complete and absolute devotion.
>
> Yours,
Amerigo Dumini.

In those years, which I remember well because I was fifteen in 1924 and there was much talk about Matteotti's murder in our home, the streets on the outskirts of town echoed in the evenings to the brazen and obscene chants of the Fascist gang-members: "The gang of assassins are we! Long live Dumini! Long live Dumini!" The chants would continue with paeans to the Duce and threats to his enemies, promising them the same fate as Matteotti. The terror felt by Mussolini and his followers diminished, however, when they realized that the outrage caused by the murder would not, as they first feared, lead to a popular uprising and the fall of fascism. Seeing that his opponents were too hesitant and unprepared, and the king too vile and inept to take advantage of what was a god-given moment to demand the resignation of his prime minister, Mussolini took the reins back in hand. Six months after the murder, on 3 January 1925, he established his dictatorship—a decisive step for which he would ultimately pay with his abject death. Counting on the tacit backing of the king and his parliamentary majority, and his tone once again imperious, Mussolini declared that, in his capacity as head of state, he was assuming all responsibility for the murder of 10 June. During the 12 January session of Parliament, the minister of the interior, Luigi Federzoni, presented a draft law for the king to hand over to the government the power to amend laws relating to internal security. On 20 May, the House was presented with a report prepared by a commission made up of nine deputies in favor of passing the bill. "Rarely," Matteotti comments, "has a report on such a bill been drawn up with such nonchalant hypocrisy in its historico-juridical arguments, with such impudent terminology, and with such a capacity to mystify by presenting problems and solution in terms of concepts so coarsely Manichean and full of pretext. Lying was becoming the science of government." On 6 November 1925, there followed the decree that led to the disbanding of opposition parties, to the suppression of the free press, to the elimination of parliamentary minorities, all under the guise of legal provisions. Thus began the repression of all those manifestations of cultural and intellectual life detested by the Fascist hierarchy and its followers, who were totally lacking in culture and always alert to the danger of transgression. The luggage of a friend of mine, who had a reputation as a dissident, was minutely inspected at a frontier. The eyes of the zealous officer on duty fell upon the sorrowful line of Giacomo Leopardi, the most famous Italian poet of the nineteenth century: *"ma la gloria non vedo"* (but I see no glory). "So," the agent said to the traveler, holding him responsible for the blasphemy, "you don't see glory. We'll see to it your sight improves."

With the self-exile, almost always clandestine, of dissidents, with the imprisonment or the house arrest of all those who refused to adhere to the impositions of the Fascist regime, Italy was deprived of her outstanding political and cultural personalities. Great numbers of the exiles emigrated to France; and many of these, with a force and tenacity prompted by despair, maintained strong ties with those who remained in Italy: their letters and articles were filtered through prison bars, were delivered to places of forced residence, went from the hands of the free into the hands of those under close watch, reviving their spirit. Among the refugees were such venerable leaders as Filippo Turati, Francesco Saverio Nitti, Carlo and Nello Rosselli (these two brothers were assassinated by order of Mussolini and his son-in-law Ciano on 9 June 1937, in Normandy, by killers belonging to the Cagoule, a secret association of the French extreme right), Claudio Treves, Gaetano Salvemini, and hundreds of others of the old and the new Resistance.

Among the most significant episodes following upon the establishment of the regime and the suppression of freedom, I recall one, not least because I followed its evolution hour by hour. This episode offered the opportunity for a first stirring of anti-Semitic sentiment. In Levi's department, a classmate of mine—though he studied biology, not medicine—was Sion Segre, a follower of the Rosselli brothers' movement Justice and Freedom. On 11 March 1934, Segre was caught by customs officers on the Swiss border as he was re-entering the country with one of Giuseppe Levi's sons, Mario, and carrying anti-Fascist leaflets and pamphlets. Sion was arrested, but Mario dived into the river and swam across, in spite of the freezing temperature that wintry day, shouting, "Long live free Italy!" The papers, however, said that he had yelled, "Cowardly Italian dogs!" The newspapers also reported the names of the fifteen persons who had been arrested as a result of the documents found on Segre, all of whom were from Turin and for the most part Jews. One of them, Leo Levi, the nephew of the rabbi of Turin, was famous in anti-Fascist circles for his tenacious opposition to the regime: "A first-rate young man," people said of him and, paraphrasing the common Italian saying "His life is all Family and Work," added, "He's all Home and Prison."

Giuseppe Levi, imprisoned that same evening in ignorance that his son Mario had succeeded in reaching Switzerland (where he lay in hospital seriously ill with pneumonia after the daring dive into the freezing river), immediately accused himself of sole responsibility for what appeared to be a conspiracy. His wife, Lidia, his children, and the in-laws who were allowed to visit him briefly tried desperately but in vain to make him understand that Mario was safe and that there was therefore no need to keep on declaring himself guilty of devising a plot of whose existence he had not even known. The policemen, though Fascist, had retained the sense of humor characteristic of Italians and painfully deficient in the German Nazis, and at a certain point decided to free the old madman, realizing that despite his claims he was in reality innocent and harmless. I remember him upon his return home, pale, tired, but happy to know that Mario was safe. As for Sion, nothing really serious had been found against him, so after a few months he returned to the laboratory. Upon his return, Salvador Luria commented, "We were living next to a historical figure and hadn't even realized it."

The seed of anti-Semitism, which—though producing small growths here and there across the peninsula—had been slow to flourish, found excellent fertilizer in the events of 1934, which were exploited with a fair amount of success by professional anti-Semites such as Giovanni Preziosi, whom even the Duce disdained, and by dozens of other such individuals. Four years were to pass, however, before the seed could find fertile ground, and a tolerant country of liberal tendency, as Italy had been at the end of the nineteenth century and the beginning of the twentieth, was to change, as a result of the actions of a few thousand vicious and corrupt individuals, into the servile vassal of Nazi Germany. In opening its doors to the barbarians, Italy was to stain itself with the same guilt. The shame that befell the nation would be lightened only through the heroic behavior of the many who rebelled.

CHAPTER 9

The Start of the Anti-Semitic Campaign: The Racial Manifesto

THE anti-Semitic campaign began in the spring of 1936: underhanded at first, with sporadic and then more frequent attacks in the newspapers; and again with renewed frequency in the first six months of 1937, dying down later in the year. When unleashed again in January 1938, it was clearly racist; and all the newspapers—from the notoriously anti-Semitic ones such as *Il Tevere, Il Regime Fascista,* and *Il Quadrivio* to those that had been neutral up till then—were increasingly strident. This change provoked the scorn of even such a long-term and committed Fascist as Emilio De Bono, who wrote in his diary: "The press is more vilely disgusting than usual." Though the few professional anti-Semites were led by two individuals held in contempt by Mussolini himself—Giovanni Preziosi and Telesio Interlandi—the number of those who joined their ranks out of political opportunism continued to grow with the hardening of the regime's attitude and the increasing closeness of the alliance with Nazi Germany. Notable among these opportunists were Roberto Farinacci, the "Ras"* of Cremona and leader of the vigilantes in the early days of fascism; the party secretary, Achille Starace, whom De Bono defined in 1940 as "the sinister buffoon"; and a growing number of journalists. Certain names have remained stamped on my memory because of the fierceness of their anti-Semitic articles which I read with ever-increasing disbelief and bitterness in newspapers such as *La Stampa,* until then considered liberal.

*A term for Ethiopian chiefs, or Rastafarians, used in Italy to identify despotic local authorities.

My initial stupor eventually gave way to a sense of liberation from the nightmare that had tormented me since early childhood when my father had mentioned the restrictions imposed on the small Jewish communities before the emancipation of the previous century had taken place, in its uneven way, in the regions of the Kingdom of Italy still awaiting unification. Even earlier, in Piedmont as in the whole of northern Italy, the breath of freedom accompanying the invasion of the Napoleonic army had been followed, early in the nineteenth century, by the restoration and reinstitution of restrictive norms; these were gradually eliminated after the unification of the kingdom under Vittorio Emanuele II in March 1861. Two years previous to unification in 1870, Piedmont had adopted a statute conferring equal rights on all citizens regardless of religion. From Father, I had also learned of persecutions still taking place and of pogroms that spread terror and death among the Jews of czarist Russia and of Poland.

The monster of anti-Semitism, all the more menacing for being invisible and yet ever present, had come out of its lair; the ungraspable phantom had become an actual and tangible reality. This fact diminished the sense of nightmare evoked during my childhood and adolescence by tales of the tribulations suffered by Jews. For the first time, I felt pride in being Jewish and not Israelite, as we had customarily been called in the liberal climate of my early years; and, though still profoundly secular, I felt a bond with those who were, like me, the victims of the lurid campaign unleashed by the Fascist press. The historian Renzo De Felice writes:

> Upon the altar of his alliance with Hitler, Mussolini sacrificed the Italian Jews without second thought even though he did not really believe in their "guilt"; thus committing a crime more monstrous even than that committed by the Nazis who at least did believe in the "guilt" of the Jews: in the same way that of his own accord he sacrificed the Jews to Hitler, had he been the ally of Stalin he would have sacrificed something else. Hitler's ally became the anti-Semite.

Edvige, Mussolini's sister, has borne witness to his cynicism in her verbatim record of an admission made to her:

> If circumstances had led me to a Rome–Moscow axis instead of a Rome–Berlin one, perhaps I would have fed Italian workers . . . the lie of the

The Start of the Anti-Semitic Campaign

Stakhanovist ethic and of the happiness it harbors. And in this case also it would have been a showy but inexpensive token of alliance.

Thinking back on the reaction of both myself and those who shared our fate, on our scorn for the vile attacks on a minority who were not only innocent of the crimes of which they were accused but incapable of defending themselves, I see a contrast between our naïve surprise and sense of rebellion and the view left by a young Dutch Jewess. Etty Hillesum disappeared in the winter of 1943 in the inferno of Auschwitz, of whose existence she was—unlike us—perfectly aware:

> My acceptance [she wrote in her diary in July 1942, when she was already aware of what awaited her] is not resignation or lack of will; there is still room for elementary moral outrage against a regime which treats human beings in this way. But the things that are happening to us are too big, too diabolical for one to react with personal rancor and bitterness. It would be a puerile reaction, out of proportion to the fatefulness of these events.

The reaction in Italy had aspects both pathetic and tragic. Representative of the first, and regarded as vile and shameful by the majority of their co-religionists, was the attempt by Jews who called themselves Fascists (and had been assimilated and belonged to the Fascist party like the vast majority of other Italians) to disclaim responsibility for crimes they claimed had not been committed, by proclaiming in a little newspaper called *La Nostra Bandiera (Our Flag)* their dissociation from the anti-Fascists and their fidelity to the regime. The publication of this newspaper was considered at the time—and is today regarded by those who blush to recall the articles of its editor-in-chief, Ettore Ovazza—an ineradicable stigma on the Turin community. De Felice, more serene and impartial, though recognizing the naïveté of the undertaking, notes:

> In regard to practically none of them can one speak of corrupt or unscrupulous people. What was at play in some of them was at most a certain dose of fear and reverential obsequiousness before constituted authority. The majority, however, were only convinced Fascists and deeply integrated (but not apostate) Jews who were as a result relatively deaf to certain more deeply moral-cultural instances of Zionism (which they opposed).

I remember the episode because of my own direct experience and the scorn my family expressed for *La Nostra Bandiera*. Ovazza in the end had to revise his beliefs in the most tragic of ways. A spy who had offered to help his son cross the Alps into Switzerland killed the young man instead, stealing all his possessions, and then betrayed his family, who were hiding in the mountains. It was then November 1943, after the Nazi takeover of the country, and informers received a reward of five thousand lire for every Jew they turned in. Ettore Ovazza, his wife, Nella, and their fifteen-year-old daughter were taken inside the village school where, according to some, they were shot, and, according to others, thrown alive in the furnace used to heat the building.

Among tragic reactions, I recall the suicide of people who enjoyed general esteem. There was the well-known publisher Angelo Formiggini, who threw himself off the Ghirlandina tower in Modena in November 1938. He left a note saying that he had chosen this way to show how absurdly wicked the racial laws were: "By suppressing myself I free my beloved family from my presence: it once again becomes pure Aryan and will remain undisturbed." He belonged, in fact, to a family many branches of which had been Catholic for several generations; and he himself was deeply assimilated. Achille Starace, the Fascist party secretary, commented: "He really died like a Jew, he threw himself off a tower to save a bullet." In May 1945, Starace himself was spared the waste of a bullet, the partisans executing him in front of the corpse of his idol Mussolini in the Piazza Loreto in Milan.

A very few of the persecuted—I was not among them, and friends who knew me at the time recall what they considered my dangerous outrage and impetuousness—reacted with serene and wise detachment. I was helped by such a person one warm autumn afternoon in 1938, shortly after the promulgation of the racial laws. Deeply saddened by the impossibility of pursuing the research that excited me at the Neurological Clinic, I went with Paola to visit a dear friend, Gina Fubini. A childhood companion, Gina was the same age as our older sister, and like a fourth sibling to us. She was being treated in a clinic for a serious pulmonary lesion. Her beautiful emaciated face with its fine features was lit by a smile very much in contrast to our anguished expressions. We found her in a wheelchair wrapped in a little shawl, enjoying the last rays of the setting sun. I asked her how she managed to preserve her serenity in face of the cataclysm that had descended upon us with such violence. She smiled at my question. "For reasons I'm not capable of explaining myself," she answered, "I feel such peace within that no force in the world could

The Start of the Anti-Semitic Campaign

affect it. I don't know whether to define it as Jewish, Catholic, or beyond religion. I know I am ready for everything and that I fear nothing."*

Gina's attitude was a gift, and one that helped me to bear the approaching storm. As for Gina, five years later she miraculously escaped capture with her husband. Having been sent to the front during the First World War, he had shown ardor and courage unexpected in a person of his sweet temper, and had then suffered the rest of his life from pulmonary lesions caused by poisonous gas. The small family—father, mother, a little girl, and a boy who could move only with great difficulty as the result of polio in infancy—had crossed the Alps in the depths of winter, and owed their safety to a snowstorm that erased their footprints. They persuaded the Swiss border guards to forgo their usual peremptory insistence on immediate repatriation, which would have let them fall straight into Nazi hands.

On 14 July 1938, there appeared in all the newspapers a racist manifesto signed by ten "Italian scientists." Only two of them enjoyed a certain reputation: the physiologist Sabato Visco and the endocrinologist Nicola Pende. The rest were biologists and doctors of no distinction and, for the most part, young and unknown assistants who wanted to get into the limelight and advance their careers. In the manifesto of the "racial scientists"—which, it was said, had been written or at least conceived by Mussolini—it was declared that the Jews did not belong to the Italian race: "Of the Semites who in the course of centuries have landed on the sacred soil of our Fatherland, nothing on the whole has remained." Among the signatories, whose names were made known only ten days after the document's publication, the two best known, Visco and Pende, protested that the text did not correspond to what they had written. They were silenced—nor is it to their merit that they acquiesced to a censorship accompanied by threats of blackmail and reprisals.

The manifesto marked the beginning of a series of anti-Semitic

---

*Etty Hillesum inscribed in her diary the same thought one June morning in 1942, before she was interned in the concentration camp at Westerbork, the antechamber to Auschwitz:

A future peace will be truly such only if everyone has first found it within themselves—if every man has freed himself of hatred for his fellows of whatever race or nation, and transformed it into something different, in the end into love, if that isn't asking too much. . . . That little piece of eternity which we carry within us can be expressed in one word or in ten volumes. I am a happy person and I celebrate this life. I celebrate it in the year of the Lord 1942, the umpteenth year of war.

provisions culminating in the decree of 17 November 1938, which forbade marriage between Aryan and Jewish citizens and deprived the latter of all the rights enjoyed by other Italians, such as that of exercising a profession, attending or engaging in didactic activity in state schools, the belonging to state firms or institutions, and so on. On 16 October of that year, on the basis of a decree previously issued on 5 September, I and all those who, like me, were declared to be of the Jewish race were suspended from academic work and from all posts in the universities and academies. Even though these provisions did not yet contain the negation of the right to life, they deprived each of us of the right to engage in any activity whatsoever in the life of the country.

I was at the time engaged, with Fabio Visintini, ordinary assistant in Turin's Clinic for Nervous and Mental Diseases, in a research project that greatly excited us and was going ahead with remarkable success, combining my competence in neurology with his in neurophysiology. Visintini had set up an oscillograph for us to record the spontaneous activity of, and stimulate by means of implanted electrodes, the nervous centers of chick embryos from the very first days of their development to later periods of ontogenesis. At the same time, using a silver impregnation technique which I had modified from Ramón y Cajal's original (see pp. 126–28), I was studying the differentiation of the neuron centers and circuits in charge of motility and reception of stimuli from spinal and medulla oblongata segments of embryos examined in serial sections under the microscope. An article on the results of these studies was presented to Italian scientific journals but was rejected because they were no longer allowed to publish articles by anyone who was Jewish. I still regard this article, which was accepted a year later by an excellent Swiss periodical, as one of my most rewarding, because of the accuracy of the analysis performed, in parallel, at the morphological, electrophysiological, and behavioral levels.

Our lab assistant at the time was a young man from the south of Italy who displayed violent adversity to the regime and its anti-Semitic measures. Unfortunately, such a high percentage of OVRA (the Fascist secret police) informers and provocateurs had infiltrated at all levels that we mistrusted him; even today I do not know whether our suspicions were correct or mistaken. Among the regime's repressive maneuverings, this universal suspicion polluting all rela-

The Start of the Anti-Semitic Campaign

tionships, which is common to all totalitarian systems, was one of the very worst. We carried out our research in an atmosphere every day more menacing as a result of the virulent newspaper articles saluting the racial campaign for opening the eyes of Italians, blind till then to the dangers of mixed marriage. In their natural wisdom the Italians found the anti-Semitic campaign repugnant; and at the academic level, I remember not a single manifestation of hostility toward me but only general demonstrations of sympathy. Nevertheless, eugenics, the science whose object is the improvement of the race, had come into fashion. One of the signatories of the 14 July manifesto had assigned to one of his students, as a topic for a doctoral thesis, the task of demonstrating the grave risk involved in marriages between Aryans and Jews. If—as the professor, who claimed to be a geneticist, affirmed—the offspring of any such marriage inherited the strong skeleton of the Aryan father and the frail organs of the Jewish mother, the organs of the halfbreed, as those of mixed parentage were then called, would be incongruous and ill adapted to the supporting frame. Even more difficult would it be for Aryan organs to be compressed within a small Jewish skeleton. So was reborn, in the twentieth century, the myth of Procrustes, the perverse thief of Greek myth who laid stray travelers on a bed and then, to make them fit it, either stretched them by force or shortened them by amputating their lower extremities. According to the myth, the slaughter ended when he was killed by Theseus. Italian eugenics ended in turn, thanks to the Theseus-like measures of the partisans and Allies.

By March 1939, I could no longer attend university institutes without risking denunciation or endangering Aryan friends, who would have become guilty of works of pietism—the accusation brought by the regime against those who manifested sympathy or solidarity with Jews. I accepted, therefore, an invitation from Dr. Laruelle, director of a neurological institute in Brussels, to continue my work at his center. The invitation was attractive both because Professor Levi had moved to Liège, having been invited by Professor Firquet to continue his scientific work in the university there, and because my sister Nina, with her husband and three children, had chosen Brussels as a temporary refuge while waiting to return to Italy or continue on to the United States.

Of the period spent in Brussels from March to December 1939, I remember the frenetic and enthusiastic activity of Laruelle's assist-

85

ant, a Rubenesque blonde we knew as Mlle. Reumont. She performed demolition work with considerable expertise on unfortunate foxes, hunted down in the woods around the city, in order to study the results of destruction of the intestinal tracts on the sympathetic system whose function was the innervation of the missing parts of the digestive canal. One day she arrived at the Institute in a state of despair, announcing: *"On a volé mon vélo!"* I thought it was a cryptic crossword clue until my friend Henny, a German, more acquainted than I with the language, explained that her bicycle had been stolen. I remember the gusting wind which swept the sky clean, the sudden bursts of rain, the sense of anxiety and fear of the population who rightly thought themselves in danger of an imminent German invasion—a danger all the more fearful as the sufferings of the First World War were still alive in memory. And finally, among my most vivid recollections, despite its insignificance, was the graffito scrawled in capital letters on the walls of certain buildings: *"Aucun café n'est bon, sans chicoré Capon."* I hated Belgian coffee and *"la chicoré Capon,"* and I hated them even more after I was told—though I don't know how credible the information was—that it was a spy's signal indicating to the Germans the town's nerve centers.

On weekends I would go to Liège to see Levi who had, with his indomitable energy, set up a tissue-culture center there. He was surrounded by students. First among these was the loyal young Chèvremont couple, who was then beginning to study various lines of cells which would lead to the recognition of the presence of DNA in mitochondria and the beginning of a new branch of genetics. Levi had resumed the investigations he had been forced to interrupt in Turin and begun new ones on the muscular system. On trains packed with young women on their way to visit fiancés and husbands posted to Liège, the most important center near the Belgian-German frontier, I was often asked: *"Vous aussi vous allez voir votre soldat?"** I would answer affirmatively—without specifying, however, that *mon soldat* was almost seventy years old.

Levi would welcome me in his gruff voice, pleased with my visit which interrupted the monotony of his days, and together we went for long walks along the avenues of Liège crowded with soldiers, women, and children, who, though living with the nightmare of

*"Are you, too, going to see your soldier?"

The Start of the Anti-Semitic Campaign

threatened invasion, were still able to enjoy their Sundays together. At the end of August, I left for Sweden to take part in a conference. On 12 September, in Stockholm, we were surprised and horrified by the news of the invasion of Poland. It was the beginning of the Second World War. Aboard one of the last ships taking civilians, I returned to Brussels. Later, in December, foreseeing the imminence of the German invasion, I decided to return to Italy, along with my sister Nina and her family, by car across a France mobilizing for war. By Christmas Eve, we were back in Italy, where Mussolini's declaration of nonbelligerence caused the naïve—including ourselves and most Italians—to hope that Italy's alliance with Germany would be broken. This hope was blasted, however, with the signing in May 1939 of the Pact of Steel, as the military alliance between Hitler and Mussolini was known.

CHAPTER 10

A Private Laboratory
à la Robinson Crusoe

SINCE I could no longer attend any university institutes, I decided at the end of December 1939 to practice medicine—in clandestine fashion, since this, too, was forbidden. I would look after those patients I had had in my care in past years when they were hospitalized in the university's Medical Clinic. These poor people, who lived in the attics of houses in old Turin, did not care about the laws and were glad of my visits and the help I could offer within the limits of my scarce finances. The constraint of having to turn to Aryan doctors to have prescriptions signed, however, forced me in spite of myself to reduce and then abandon this activity. I took refuge in reading and cultivated relationships with the many friends who scorned the danger of being accused of pietism.

On 10 June 1940, a dear friend and I were hard at work writing her doctoral thesis when, alarmed by the unusual activity in the streets, we opened the windows. It was six o'clock in the evening. From the loudspeakers set up in the piazzas and main thoroughfares came the loud voice of the Duce: "Fighting men of the army, of the sea and of the air. Blackshirts of the revolution and of the legions. . . . Attention! A fateful hour is striking in the sky of our Fatherland. The hour of irrevocable decisions. A declaration of war has already been presented to the ambassadors." From the Piazza Venezia, where an immense crowd had been assembled, the listeners still ignorant of the ambassadors in question roared their enthusiastic approval. Once the patriotic fervor orchestrated by the Fascist leaders had died down, he continued, stressing each word: " . . . to the ambassadors of Great Britain and France."

A Private Laboratory *à la* Robinson Crusoe

Italy's ignoble attack on France, by then already in extremis, started the next day. The "Battle of the Alps" cost the French an exiguous number of casualties, the Italians about two thousand, and many cases of frostbite because the summer equipment of the latter proved inadequate to a sudden onslaught of winter cold. In the months that followed, Italy's lack of military preparation was revealed in all its tragedy on the Greek and East African fronts, while Naples and Sicily were bombed intensively by the British air force.

A few months after the beginning of the war, in the fall of 1940, Rodolfo Amprino, recently returned from the States, came to visit me. I was surprised when he asked me, in a brusque, "Piedmontese" manner, about my projects, it not having occurred to me, in that wartime climate, that my personal problems would be anything but irrelevant to others. My surprise was even greater in that my relations with Rodolfo, since our meeting eight years before at the Anatomy Institute, had been limited to laconic exchanges of information on histological techniques. Or rather, to be more precise, on the basis of his mastery in the field he would tell me things and I would clumsily follow his instructions.

My silence in response to his question provoked a sudden and somewhat irritated reaction: "One doesn't lose heart in the face of the first difficulties. Set up a small laboratory and take up your interrupted research. Remember Ramón y Cajal who in a poorly equipped institute, in the sleepy city that Valencia must have been in the middle of the last century, did the fundamental work that established the basis of all we know about the nervous system of vertebrates." Rodolfo could not have sown his suggestion on more fertile ground. At that moment he seemed to me Ulysses as Dante immortalizes him in the twenty-sixth canto of the *Inferno,* when the Greek hero encourages his fellow voyagers not to lose heart but to continue on their course toward the unknown: "My companions I made so eager for the road with these brief words that then I could hardly have held them back."

Rodolfo had, in fact, touched a chord that had been vibrating in me since earliest childhood: the desire to undertake a voyage of adventure to unknown lands. Even more appealing than virgin forest was the jungle lying before me at that moment: the nervous system, with its billions of cells gathered in populations each different from the other and all locked into the apparently inextricable nets of the nervous circuits which intersect in all directions along the cerebrospinal axis. The pleasure I was already savoring in anticipation was enhanced by the prospect of carrying out the project under the conditions contingent on the prohibitive racial laws. If Ramón y Cajal, with his giant's step and exceptional intuition, had dared foray into that jungle, why should I not venture along the path he had opened for me? My first experience with Visintini had been encouraging. Though unable to continue the same line of research, lacking both space (I only had my small bedroom to work in) and competence in electrophysiology, I could nonetheless analyze other aspects of the developing nervous system by relying on my expertise in the selective coloring of nervous tissues with the silver-impregnation technique and on my ability in microsurgery. Ramón y Cajal's success—as well as the much more modest one of Visintini and myself in studying the function and structure of the nervous system of chick embryos—was the result of the tactic of studying the system in its *statu nascendi*, when it is made up of only some thousand cells interconnected by a still moderate number of neuronal circuits. Chick embryos were ideal material also because they could be easily procured and incubated at home.

I submitted my intention to Mother, Gino, and Paola and received their approval. Mother, in fact, would have been willing to accept any sacrifice rather than face another separation. Gino and Paola, who had been against our moving to the United States because they were both strongly bound to Italy and confident that Nazi-Fascism would be defeated, understood my need to resume the work so suddenly interrupted by my return from Belgium.

The instruments necessary for the realization of this project were few. The need for an incubator was met by a small thermostat which worked very well for the purpose. Another, high-temperature one served to seal the embryos in paraffin. The embryos were then silver-stained and cut into series using a microtomer. The most expensive items were a stereomicroscope, needed for operating on the embryos, and a binocular Zeiss microscope with all the eyepieces and photographic apparatus. The equipment was completed by a series of watchmaker's forceps, ophthalmic microscissors and surgical instruments consisting of common sewing needles which, with a very fine-grained grindstone, I transformed into extremely sharp microscalpels and spatulae. The collection of instruments, glassware, and chemical reagents was like what one of my nineteenth-century predecessors would have found necessary. Gino built me a glass thermoregulated box with two circular openings on the front. Through these I could insert my arms and operate on the embryos under the microscope in an environment of 38° C., protected against possible infection: a point of caution that revealed itself to be entirely unnecessary but had the advantage of surrounding me with a religious sort of respect. My mother also saw to the maintaining of the latter by forbidding my room to curious visitors, telling them that I was operating and could not be disturbed. The "tour de force" of fitting so much apparatus in the small space available to me added to the pleasure of managing to work under prohibitive conditions. The most cumbersome piece was undoubtedly dear old Levi himself who, with his great corporeal mass and meager agility, threatened to destroy all the carefully laid out histological sections with a mere swinging of his large hands each time he moved. "Excuse me, I'll be more careful" he would mutter, without, however, giving too much weight to these accidents on the job.

I spent the winter and spring of 1941 busy with preparations and the first experiments, which turned out well. The worsening of the military situation and the defeat of the Italians in northern Africa, with the English occupation of Ethiopia in April and the loss of eastern Africa, caused the campaign against the Jews to become even more bitter: they were now enemies to be fought on the home front to make up for losses suffered beyond the country's borders. Articles in the newspapers were matched by anti-Semitic graffiti and posters pasted on walls all over the city. On 16 October, Gino came home proud of an honor paid to him. "They've put me in Einstein's company," he told us. On the poster whose contents the partisan Emanuele Artom reproduced in his diary, Gino's name is listed immediately in front of Einstein's, along with other eminent persons belonging to the "Jewish race," such as Franklin Delano Roosevelt, La Pasionaria, Haile Selassie, and Lenin. After listing the horrible crimes committed by these "Jews," the manifesto urged their punishment: "Are we going to put an end to it once and for all, then? Not to the concentration camps, but up against the wall and then at them with a flame thrower! Long live the Duce! Long live Hitler!" The following day, vaguely aware of the fact that they had included in their list the names of persons who did not belong to the Jewish race, the authors urged the citizenry to fire upon Jews at the slightest suspicion, entrusting the Creator with the posthumous task of discriminating any errors. Artom comments in his diary: "To read posters in which one is threatened with death, accused of many crimes, is an experience it is not given to everybody to endure." He was to endure it to the last drop. Revealed early in 1944 as the political commissar of the Action Party, a Jew, and a partisan, he was subjected to horrible tortures at the hands of the SS and killed without his torturers being able to get a single name or complaint out of him. All this his mother learned from Oscar, another partisan who witnessed Emanuele's death.

In 1941, and up to the time of the Nazi invasion of the country, insults and threats were not followed by acts of actual persecution. Thus, from the spring of 1941, I was able to carry on, in the calm of a minuscule laboratory not unlike a convent cell, a research problem that absorbed all of my time from then until the invasion. My aim was to analyze how excision of still non-innervated tissues in the peripheral territories, or limbs, affects the differentiation and subsequent development both of motor cells in the spinal cord and of sensory cells in the dorsal root ganglia at a very early stage of embryonic life. Previously, this problem, one of the first to be tackled experimentally by researchers into the development of the nervous system in the first two decades of the century, had scarcely interested me. Amphibian tadpoles had been the subject of these experiments, but the results obtained seemed to me too vague to lend themselves to satisfactory interpretation. These findings suddenly appeared in a different light one summer day of 1940, shortly after Italy had entered the war on Germany's side.

My conversion, if such it can be called, occurred while I was riding on a train used before the war for the transportation of livestock. After war had been declared, civilian trains were taken over for troop transportation, and these livestock trains, or cattle cars, were used for civilians, for short journeys in the provinces. The wagons, which lacked seats, doors, and windows, offered great panoramic views through the windowless open sides. I was traveling on one, along with my friend Guido, the formidable whistler of classical arias, on our way to a small mountain village. I sat down in what was considered to be one of the best places—the floor of the wagon—with my legs dangling over the side in the open air. The slow progress of the train, the vertical bars that offered firm support

A Private Laboratory à la Robinson Crusoe

and, in my case, Guido's vigilant hand, ensured against possible falls and allowed me to see the fields in their full summer's growth. While enjoying the view and the air which smelled of hay, I was distractedly reading an article Levi had given to me two years before. Published in 1934 in an American periodical, it was the work of a pupil of Hans Spemann, the German biologist who was awarded the Nobel Prize in 1935 for his discovery of a factor (or of factors, which even today have not been precisely identified) called the "organizer" because of its property of inducing the differentiation of organs and of whole embryos that come into direct contact with the tissues releasing it. The author of the article, Viktor Hamburger, had not analyzed this phenomenon, but his interpretation of the effect he described was clearly influenced by the same concept of an inductive reaction of certain tissues on others during the early stages of embryonic development. Hamburger had studied how the ablation of chick embryo limb buds affected the sensory and motor neurons responsible for their innervation. The author observed that, one week after such an operation, the motor column and the sensory spinal ganglia responsible for the innervation of the limbs were greatly reduced in volume. Hamburger interpreted these findings as pointing to the absence of an inductive factor, otherwise normally released by the innervated tissues and necessary for the differentiation of motor and sensory nerve cells; without the factor, these cells could not undergo differentiation. For me, Hamburger's limpid style and the rigor of his analysis—in sharp contrast with those of previous authors who had described the same phenomenon in amphibian larvae—cast new light on the problem. I don't know how far the idyllic circumstances in which I read the article contributed to my desire to delve into this phenomenon, but in memory my decision is indissolubly bound up with that summer afternoon and the smell of hay wafting into the wagon. I did not imagine at the time, however, that this interest and my subsequent research would determine my future.

The summer of 1941 was overshadowed by the anguish caused by the news of the triumphant advance of Hitler's troops into Russia and of their successes on all fronts. We were, furthermore, deprived of all news of Levi whom we feared had fallen into Nazi hands since his refusal to leave Belgium after the German occupation. It was with immense joy that, at summer's end, we welcomed him back— shockingly thin and pale after a dangerous trip across Germany. It turned out that, after a year in Nazi-controlled Liège, meeting his friends secretly in a little café outside of town, he had been unable to endure any longer the hunger, loneliness, and boredom, and had set out for home. I was especially happy to see him, and asked him to join me in my new research. He accepted with great pleasure, and thus it was that from that autumn till a year later when we were both forced to leave the city, his imperious voice resounded in my bedroom-laboratory from morning to night. Work would be interrupted when his loyal pupils arrived, and the topic of conversation would shift from chick embryos to the madmen and criminals who were running our country.

In the winter and spring of 1942, our research yielded unhoped-for successes. The examination of embryos whose budding limbs had been excised in three-day specimens, and impregnated using the silver technique, revealed with extraordinary clarity the nerve cells and the fibers that sprang both out of the motor neurons of the spinal column and out of the sensory ganglia in embryos sacrificed at brief intervals from the time of the excision to the end of the twenty-day incubation period. These findings suggested a different explanation from the one advanced by Hamburger to explain the almost complete disappearance of the motor cells in the spinal ganglia that innervated the limbs in embryos not subjected to such destructive treatment. It was a question of the absence not of an inductive factor necessary to their differentiation, but of a trophic factor that is released by innervated tissues and that, under normal conditions, the nerve fibers convey toward the cellular bodies. In fact, in embryos with excised limbs, the differentiation of nerve cells proceeds normally, but a degenerative process followed by the death of the cells begins to occur as soon as the fibers springing out of the cord and from the ganglia reach the stump of the amputated limb. Their death appeared to be caused by the absence of a trophic factor and not, as Hamburger had hypothesized, by an inductive one belonging to the category of those known as "organizers."

Many years later, I often asked myself how we could have dedicated ourselves with such enthusiasm to solving this small neuroembryological problem while German armies were advancing throughout Europe, spreading destruction and death wherever they went and threatening the very survival of Western civilization. The answer lies in the desperate and partially unconscious desire of human beings to

A Private Laboratory *à la* Robinson Crusoe

ignore what is happening in situations where full awareness might lead one to self-destruction.

In the second half of 1942, with the Allies' systematic bombing of the cities in northern Italy and of Turin in particular, a favorite target because of its great industries, life in the city became every day more dangerous. Almost every night, the lugubrious whine of sirens, warning of British planes overhead, forced us to go down into the basement in spite of the risk—which became tragic reality for hundreds of people—of being buried under the ruins of bombed buildings. Every time the alarm sounded, I would carry down to the precarious safety of the cellars the Zeiss binocular microscope and my most precious silver-stained embryonic sections. These vigils usually dragged on for hours, amid the murmurs of women praying, until the sirens announced that the danger had, for the time being, come to an end. Very often, however, a squeal announcing a new series of planes forced us to rush down again.

When autumn was well under way, we decided, like the majority of the people of Turin, to move out of town. Thus, in a small house in the hilly Astigiano highlands, an hour away from Turin, I set up my laboratory on a small table in the corner of a room that served also as dining area and family sitting room. Since eggs had become extremely scarce, I cycled from one hill to another begging farmers to sell me some "for my babies." Casually I inquired whether there were roosters in the chicken coop because, as I explained, "fertilized eggs are more nutritious." A unforeseen difficulty arose when my activities in the common room fell under Gino's eyes. He noticed how I used spatulae and ophthalmic scissors to extract from the eggs five-day-old embryos that had been operated on, and how then, instead of throwing the eggs away, I carried them into the kitchen and used them to fix our meals. From that day on, he categorically refused to eat the scrambled eggs and omelets, which up until then he had thought excellent.

Levi lived in a different locality outside of Turin but returned every day. On alternate days, and always after heavy bombings, which my family and I would witness in dismay from the top of the hill, gazing at the sky lit by the glare of fires, I also returned to Turin to meet him. With other friends, we warmed ourselves by a stove in the kitchen of my home, the only warm room in the freezing apartment, and ate the steaming cornmeal that an old housekeeper poured us from the pot while recounting the night's events. With a smiling, rubicund face, he expressed the pride he took in his work, saying, "I do everything on my own"—though doing nothing but stir the cornmeal. These were the most serene moments of those wintry days in the city devastated by night bombing raids. The ruins of bombed buildings, broken pipelines, damaged electrical and telephone plants were swept aside and repaired with unbelievable speed, but hopelessness and despair were written on everybody's face. At dusk began the assault on the overcrowded trains carrying people back to the shelters scattered about in the hills.

In spite of the almost prohibitive conditions—the difficulty of procuring fertilized eggs, and the repeated failure of the power supply upon which depended the functioning of my incubator and the development of the embryos—I completed some projects which I was to carry further a few years later in the United States. Their central theme was the study of the interaction of genetic and environmental factors in the regulation of the differentiation processes of the nervous system during the early stages of its development. At the beginning of spring, from the window of my small room in our cottage, I contemplated the ducklings following their mother in single file, diving from time to time as she did into the ditches that, after rain, flowed down the sides of the little road I used to cycle along every day. In specific areas of the embryonic nervous system, cells in the first stages of differentiation detach themselves from cellular clusters of cephalic nuclei and move singularly, one after the other like the little ducklings, toward distant locations along rigidly programmed routes, as is demonstrated by the fact that the spatial and temporal modalities of these migrations are identical in different embryos. In other sectors of the developing nervous system, thousands of cells move about like colonies of migrating birds or insects—like the Biblical locusts that I was to see many years later in Ecuador. The fact that I was for the first time observing natural phenomena unknown to those who live in cities, such as the springtime awakening of nature, cheered me and stimulated my interest in studying the developing nervous system. Now the nervous system appeared to me in a different light from its description in textbooks of neuroanatomy, where its structure is described as rigid and unchangeable. Only by following, from hour to hour in different specimens, as in a cinematographic sequence, the development of nerve centers and circuits, did I come to realize how dynamic these processes are; how individual cells behave in a way similar to that of living beings; how plastic and malleable is the entire nervous system. This system, which more than any other must adapt its structure and functions to environmental requirements, was to remain the main object of my research in the years that followed. Its analysis came into focus and grew in that country milieu probably much better than it would have in an academic institution.

In the summer of 1943 occurred the event marking the end of an era for Italy and the beginning of a most dramatic period. On the evening of 25 July, at 10:45 P.M., while we were listening to the radio, the program was interrupted by an announcer reading out the following news:

> Attention, Attention! His Majesty the King Emperor has accepted the resignation from the offices of Head of the Government presented by the Prime Minister and Secretary of State His Excellency Cavalier Benito Mussolini, and has nominated Head of the Government, Prime Minister and Secretary of State, His Excellency, the Marshal of Italy, Pietro Badoglio.

The news was received in my home, as throughout the entire peninsula, with immense jubilation. The demonstrations of enthusiasm were profoundly genuine yet, at the same time, indicative of a collective irresponsibility—or, to describe it in less severe terms, a lack of awareness of the danger looming over us, with German troops stationed in Italy and more massing on its frontiers.

The following morning, I went into Turin as usual. On the train, people were hugging each other, crying and laughing. At the station, from trains that until the previous day had spilled onto the platforms a gloomy and silent crowd, now descended passengers who behaved as if they were intoxicated. They began to cast off Fascist insignia— until the previous day, the precious symbols of support for the regime, and now objects of derision and shame. Leading fascists stayed shut up in their houses that day and the following ones; less important party members mixed in with the crowd who accepted them good-naturedly. Everybody, for that matter, felt somewhat guilty.

If optimism among the "Aryans" was to some extent justified, as they were not directly in the Nazis' gunsights, it was, on the contrary, completely absurd in the small Jewish population, ourselves included. Even though we were only partly aware of what had happened and was still happening in the European countries invaded by the Germans, it was folly not to have taken immediate precautions to save ourselves after 25 July. Faith in the Italians who welcomed us back among them, a common hatred for Nazism, and the absurd conviction that what happened in other countries could not happen in Italy were the source of this irresponsible attitude, an attitude which was to cause thousands of people untold suffering and death.

CHAPTER 11

Life in Hiding

IN the early summer of 1942, there had stood out, among the middle-aged and elderly women of Turin who all night long recited prayers in the basement—between the sirens' wailing to announce the approach of planes and again to inform us that the danger had passed—a remarkably beautiful young woman, Mariuccia, then barely twenty. Her life was from that time to be indissolubly linked to that of Gino. In spite of the eighteen years difference between them, hardly noticeable because of his extraordinary youthfulness, a love was born and they were married a year later. Though the simple religious ceremony had, owing to the racial laws, no legal value, it sealed the bond that was to unite them for life. Gino's life, as ours, was then in constant danger. The fact that her parents, landowners from Lombardy, gave their full consent to the wedding of their only daughter to a Jew (though Gino had years earlier become a Catholic) testifies to the scorn in which the racial laws were held, and shows that the venomous press campaign had not influenced the great majority of Italian citizens.

From the end of July 1942—the days of the heaviest bombings—Mariuccia spent many hours of the day with us in order to consult Gino—so she said—on problems of aesthetics, the topic of her doctoral thesis in literature. She then came to stay at our refuge, the small house in the Asti region; and since she taught in a high school in Turin, she and I used to go into town and return to the hills in the evening together. The wedding took place in Turin on the fateful 8 September 1943. In the afternoon, Gino and Mariuccia left for a brief honeymoon vacation, while Mother, Paola, and I returned to our mountain village, where we planned to spend a few more weeks. While on the bus, we heard the loudspeakers calling the population together. The bus stopped, and we listened with the crowd. It was six-thirty in the evening. Over the loudspeakers came a serious and gloomy message from Pietro Badoglio, new head of the government and supreme commander of the armed forces. He announced that Italian troops had ceased all hostilities against the Anglo-Americans, but would counter, with force if necessary, any enemy action *from whatever side.*

During the night of 8 and 9 September, the king, the crown prince, the rest of the royal family, along with Badoglio, the generals Vittorio Ambrosio and Mario Roatta, and their various entourages lost no time in getting to safety in what public contempt came to call the "flight from Pescara." Once they had reached the city of that name on the Adriatic coast, they proceeded south by boat to Brindisi, in Apulia on the heel of Italy which had still not been invaded by Hitler's armies. Their journey was not hindered by the Germans stationed in Italy who were, in fact, pleased to see the backs of the royal family and of Badoglio head of the armed forces. The German divisions came down the Brenner Pass and in a few hours were spreading throughout the Po valley. The Fascist generals charged with the defense of big cities like Milan and Turin handed them over to the Germans and ordered their troops to remain passive in order to "avoid useless bloodshed."

On 10 September, we saw German tanks outside Turin's central railway station, where a little over a month before I had been part of a joyous crowd throwing away Fascist lapel badges. Now people were milling around, sullen and silent, while the German military assumed control of city traffic. As far as my family was concerned (Gino and Mariuccia had returned from their honeymoon almost immediately), we knew that a delay of days, perhaps even of hours, might cost us our lives. Together with Nina and her family, we struggled to decide which path to take. Three possibilities presented themselves, each bristling with risk: to stay in the Piedmont; to try crossing into Switzerland; or to take the chance of heading south in the hope of early liberation at the hands of the British and Americans, who were by then in control of Sicily and had already landed along the southern coast of the peninsula. Mother's brother, who had been living with us, could not be dissuaded from remaining in Asti, his native city. He had been hospitalized in a clinic before 8 September and there remained, protected by nuns and doctors. Subsequently, he attributed his evasion of capture to his own naïve precaution of wearing a cap over his eyes and his lapel turned up, "so as not to be recognized," whenever he cautiously left his room. In reality, everybody in Asti knew of his presence in the clinic, but so deep was the hatred for the Nazi-Fascists, and so strong the general pity for the persecuted, that no one informed on him or on many other clandestine Jews. Indeed, in courageous solidarity, thousands of Italians offered Jews protection at grave risk to themselves.

Our first plan, as absurd as it was risky, was to seek refuge in Switzerland, as had been successfully attempted by those who had acted immediately after 8 September. By about the middle of the month, as it then was, the possibility of crossing the frontier without subterfuge was entirely out of the question. At Porto Ceresio, on Lake Lugano, where we arrived late one afternoon in the third week of the month with Nina and her family, we were greeted by the icy silence of a frontier post patrolled by a few Italian guards, and by German military personnel who fortunately failed to notice us. In the following days, Nina and her family succeeded after a thousand vicissitudes in crossing the border illegally at a less well-guarded part; but the rest of us had taken the first train back. Immediately thereafter, we set about the last of the options we had considered: taking our chances and moving south. We were helped by the fact that the trains, leaving and arriving according to no schedule, were crowded with civilians and ex-military who had thrown away their uniforms and were heading either home or toward the partisan formations scattered about the peninsula. Almost everybody, and we among them, was furnished with identity cards which had nothing in common with the real ones except sex and a plausible approximation to the right age. Our destination was the outcome of chance. In the carriage in which we were traveling, I found myself face to face with a young officer, a fellow student from university days, who recognized me and asked where we were heading. His Fascist uniform and his question increased my embarrassment: I couldn't let him in on the fact that we didn't know where we were going. I told him that in our hurry we had boarded on the wrong train and would therefore be getting off at the next station. We were happy to find that it was Santa Maria Novella in Florence. Under a heavy autumnal downpour, at six o'clock on the morning of 8 October, we got off the train with the immediate aim of cutting short a difficult conversation and the ultimate aim of finding temporary lodging for the night. A friend of

Paola's, Marisa Mori, and her mother, awakened by our early-morning phone call, welcomed us with joy. Once in their hands, our problem was immediately solved, not only for the night but for the entire duration of the war: that is, until May 1945.

Signora Consilia Leoncini, whom Marisa took us to meet, greeted us with a sad, kind smile. Her handsome face was marked by tiredness; she spent her day looking after her sick father and her sister, who was also very ill. She looked at Mother, Paola, and myself with a slight air of diffidence. "I have a room available," she said, "and would be glad to rent it on the condition that you are not Jews." She hastened to specify that she had no prejudice against Jews, having indeed many Jewish friends, but that she didn't feel like taking risks at the moment, with her father being sick and her son, from whom she had had no news for some time, away at the front. We assured her that we were not Jews but Catholics from Apulia, that the war had taken us by surprise while up north and we couldn't return home because our little town was occupied by British and American troops. Without any comment—we didn't know how far she believed us—she assigned us a large room. It was rather dark because the apartment was on the ground floor of a building on one of the city's central streets and the sun never entered it in winter. Gino and Mariuccia found lodging not far from us. A few weeks after our arrival, Signora Consilia's father and sister died, as had been expected. These deaths did not alter the course of our days which, uniform and monotonous, were interrupted only by the daily visits of Cosetta, the daughter of Signora Consilia, and her husband Ernesto, and of friends of ours who, like us, had sought refuge in Florence.

After the first few days, I had looked into the possibility of going on with my work at the university in Florence—an idea that alarmed the director of the Neurological Institute, who said that there was no way for me to attend it "incognito." So Paola and I dedicated ourselves to the only activity not listed either as prohibited or as permitted: we filled out false ID cards and distributed them to friends who had managed to reach the city of Florence. In Turin, my friend Olga Villa had given me a considerable number of these cards, printed by partisans. So obviously and, I think, brazenly false were they, however, that they would not have fooled even the most stupid German agent. The Italians, on the other hand, were well aware that these fake documents were circulating; I believe they enjoyed it, and they were careful not to ask embarrassing questions.

102

I am ashamed to recall the unforgivable mistake I made when—each person in the family being in urgent need of an ID card upon our departure from the Piedmont—I assumed the responsibility of preparing them. We had chosen two surnames—Lupani for Mother, Paola, and myself; and Locatelli for Gino and Mariuccia—lest the same surname put everyone in danger if, by some mischance, either party should fall into the hands of the Nazis. The cards, even though they lacked one of the seals, could survive a careless glance. More serious was the fact that their numbering was serial, while the dates on which they had supposedly been issued differed in the case of the "Locatellis" by about a year from our own. It was inconceivable that, even in the small village in the south given as our address, no ID cards had been issued for an entire year. Aware of this extremely serious, possibly fatal oversight, we were terrified when, like everyone else residing in the city, we were forced to show our IDs in order to obtain the card entitling one to purchase a small daily ration of food. Failure to possess this document would have been cause for suspicion, and we could not afford the risk. Paola, Mother, and I presented ourselves at the municipal offices, wondering whether we would come out again or be detained for "interrogation." The situation worsened when the clerk, having noticed that Paola and I had the same birthdate, asked Mother the obvious question: Were we twins? In the confusion of the moment and uncertain about the birthdates we had put down, Mother hesitated in answering the banal question. When the clerk returned our IDs with a smile and, with them, our ration cards, we blessed the humane "Italian-style" interpretation given to the "interrogation" municipal employees were supposed to carry out. With a pride that probably puzzled her, we showed Signora Consilia the ration cards. For us they had a far greater value than usual for such documents: they sanctioned, at least temporarily, the right to life, of which we had been deprived by the Nazi invasion of Italy.

On 11 February 1944, while I was out in the Piazza San Marco buying some rubbery, sickly sweet bread-cakes whose sole merit was the alleviation of hunger, I learned from the newspaper headlines and the excited comments of passersby that at dawn that day, on the firing range of Verona, a firing squad had executed Mussolini's son-in-law, Galeazzo Ciano, and four other members of the government who had been among the signatories of the motion demanding Mussolini's resignation, which had led to his fall and the collapse of the regime. To the news of the execution, decreed by Mussolini upon

Hitler's orders, Italians responded with indifference: partisans were dying by the thousands in the hills or being tortured in the prisons; and the five men executed were, among others, responsible for these tragedies.

Previously, we had been overjoyed to discover that Signora Consilia, Cosetta, and Ernesto were ardent anti-Fascists. Every evening we all listened to the radio, to the familiar, encouraging voice of Colonel Stevens on the BBC, and anxiously followed the news both of the battle then being fought around the monastery at Monte Cassino south of Rome, and of the slow advance of the British and American troops up the Italian peninsula. One spring day, while I was in bed with a slight illness, the bell rang, and I recognized the imperious voice of Professor Levi asking for me. Thanks to the timely intervention of his son Gino who had been warned that his father's arrest was imminent, Levi had miraculously escaped capture in the small village in Piedmont where he had taken refuge. Now he had come to Florence, where his wife Lidia and two children were staying. To the Signora Consilia, who asked whom she should announce, he answered, "Professor Levi——Ah, no, I forgot. Professor Lovisato." We thought then that his inexplicable "absent-mindedness" had caused Signora Consilia to become suspicious for the first time about our identity and that of the friends who often came to visit us. I have recently learned from Cosetta that her mother had almost immediately realized who we were, when during their daily exchanges Mother fell into repeated contradictions on the degree of her parentage with Gino and about the past. Signora Consilia had confided her suspicions to Ernesto, an able engineer known as an anti-Fascist, and to her daughter Cosetta, and they had decided to play along with us and accept grave risks rather than aggravate our situation.

After Levi came to live not far from us, we got back into the habit of going for long walks as we used to do in Liège, and of passing a lot of time together revising the rough drafts of a new edition of his two weighty volumes on histology. On 3 August, a state of emergency was proclaimed in Florence which limited freedom of movement during the day and imposed a curfew in the evening and at night. Since Levi was in Fiesole at his daughter Paola's, I managed to reach his apartment and salvage his manuscript. Electricity, water, and bread were no longer available.

That night the population was startled from sleep by the ominous sound of falling bridges. The mining of the bridges, the crowning crime committed by the Nazis in the course of their systematic destruction of all ethical and cultural values, deeply wounded the Florentines, who crowded the banks of the Arno to witness in dismay the disappearance of the marvelous Bridge of the Trinità and the other bridges over the river. From then on supplies to the city, which came almost entirely from the orchards on the right bank of the river, were virtually cut off.

At six o'clock on the morning of 11 August, the clanging of the bells of the Palazzo Vecchio announced the beginning of the insurrection organized by the Tuscan Regional Committee. From that day till 2 September, various parts of the city fell alternately into German and partisan hands. From the roofs, snipers in the pay of the Nazis and Fascists were constantly firing at passersby. Even though the war was going on outside our windows (it seemed as if the battlefront ran right down the street where we lived, so frequent was the firing of rifle and machine gun which hit civilians and military alike), we were breathing an exhilarating air of freedom and still went out into the streets heedless of the dangers. We also risked stepping on one of the mines that the Germans had scattered about and which caused many deaths. Every day we witnessed the migration of crowds of people through the streets, leaving their homes—because the Germans had come back—and carrying with them all their most precious possessions. This mass movement of the population, before the final expulsion of the Germans and the arrival of the Allied armies, became the subject of one of Paola's most beautiful paintings of that period: "The Walking City." That's how we saw it from our windows or from the sidewalks: a confused intertwining of human figures, represented in her painting as broken vertical lines, crossed arms and bent heads, a chaotic complex of people and household effects which moved without fixed destination from one part of the city to the other.

On 2 September, the British marched silently along the streets packed with onlookers. For the first time I saw, after the soldiers had passed, a bus marked with a Star of David—now no longer an object of derision. Water tanks bearing the emblem were going around distributing water to the thirsty population. In a state of mind quite different from that in which we had presented our false IDs to receive our ration cards, we returned to the same offices to present our real ones—which we had kept jealously hidden during our clandestine life—to be recognized in our real identities.

CHAPTER 12

Caring for War Refugees

WITH my Red Cross insignia, to which my degree in medicine entitled me, I was able to move about the city even during curfew hours. Thus, at the beginning of September 1944, I presented myself to the Allies' health service, and was assigned, along with three other doctors, to what was to be my most intense, most exhausting, and final experience as a medical doctor. After Florence and central Italy had been liberated, fighting between partisans and British and Americans, on the one hand, and Germans, on the other, continued in the Apennines south of Bologna, along what was to go down in history as the "Gothic line." The inhabitants of these mountainous regions between Tuscany and Emilia-Romagna were caught in the middle of the battleground. Allied trucks quartered the constantly shifting front, collecting whole families of farm workers endangered by the fighting and bringing them by the hundreds to Florence, where they were housed in temporary shelters. I had been assigned—along with Giuseppe Levi's son Alberto, a fellow student from university days— to the health service in one of the building complexes used for this purpose. The complex, an old military barrack in total decay, consisted of enormous sheds with no dividing walls, with only partitions separating what had been used as stables from the soldiers' dormitories. The earth floor was covered with straw, and the beds were straw mattresses, one next to the other. The upper floor was used as both food store and dispensary; the infirmary, where the nurses worked, was on the ground floor; there was only one small, separate room for the doctor on duty.

My forced inactivity while living in Florence having sharpened my desire to engage in some useful work, I dedicated myself with enthusiasm to the double one of doctor and nurse. The trucks always

106

arrived during the night. Anxiously I awaited their arrival and the unloading in the barrack's yard of the refugees, exhausted by the long journey and the hardships they had endured. My duty was to see that their state of health was checked and then that they were settled somewhere in the shed. The consequences of malnutrition and of the cold were particularly evident in babies and old people. I had never in the course of my brief experience as a practicing physician— before the racial laws prevented me from frequenting hospital wards—witnessed a sight so painful. Many of the newborn children arrived in a state of extreme dehydration. At dawn I would take them to the pediatric clinic, only to be an impotent witness to the extinguishing of the lives of those small creatures, by then already on the threshold of death.

In that same period, my entire family was extremely worried for Gino and Mariuccia's first child, Emanuele, born on 29 May, on the eve of the bombing of Florence. His mother's incapacity to breastfeed him sufficiently—she had to enter a hospital only a few weeks later— and the difficulty we had in procuring milk in that time of emergency, were remedied only by the arrival of the Allies and their distribution of powdered milk. However, for the little refugees from Emilia-Romagna the remedy arrived too late.

Toward the end of winter, an ever-increasing number of refugees were suffering from fever and digestive disturbances. In those overcrowded and promiscuous living conditions, the disease spread within a few days and became epidemic in character. We were horrified to realize that it was an epidemic of abdominal typhoid, a disease endemic in Italy at the time and for which, before the discovery of antibiotics only a few months later, there was no valid cure. Given the scarcity of beds in the infirmary, it was impossible to isolate the most serious cases, which soon numbered in the hundreds. A health commission led by an English woman doctor ordered the immediate transfer of those infected to a city hospital, where pyretotherapy was undertaken in the hope that provoking fever would increase resistance to the disease. The results were disastrous. News of the death of those hospitalized reached the barrack; and in a few days, the patience and faith of the refugees turned into outright rebellion whenever a transfer to hospital was ordered.

By then I was spending all hours of day and night close to the most seriously ill, in the capacity more of nurse than of doctor, our nursing service being almost nonexistent. The risk of contagion, to which I was at every moment exposed, diminished my never entirely alleviated sense of guilt for not having taken a more active part in the partisan resistance. I had been dissuaded from doing so by the mortal danger to which Mother was exposed and the imperative that she not be abandoned, as well as by what Guido had defined as my constitutional ineptitude for the conspiratorial life. It was for want not of courage but of a presence of mind without which not only my own life but that of comrades as well would have been endangered were I ever to have fallen in Nazi hands.

Now the rebellious refugees, who up until then had set all their trust in me after witnessing my dedication to their care, hid whenever it was their turn to be examined, fearing that a rise in temperature or signs of the infection, so distressingly evident, would require their hospitalization. In the worst period, the death count went up to fifty a day, and the hostility and rebelliousness of the refugees increased along with their mortality. An inquiry into the causes of the disease, which the English doctor blamed on inadequate medical inspections, showed that it was due to pollution of the refugees' drinking water. By the time the cause had been identified and eliminated, the epidemic had already resulted in carnage and thrown into despair all those who had already contracted the typhoid or felt as if they might do so at any moment.

Among the hundreds of cases I dealt with, one has remained most painfully vivid in memory: a farming couple who had a very beautiful daughter just turned twenty. They had become attached to me from the start and confided in me the difficulties and dangers they had passed through in the course of the war. One day the daughter showed the first symptoms of the infection. Her parents begged me to prevent her being sent to hospital, and I would have gladly broken the rule had I thought that there was any chance of saving her. Death struck in the barrack as it did in hospital, and her state left no room for hope, especially as she had fallen into the stupor typical of this bacterial infection. My sense of impotence in this case contributed to the decision I later took not to practice the profession. I lacked, in fact, the detachment that allows a doctor to face the suffering of a patient without creating an emotional involvement damaging to both parties.

As spring approached, the epidemic began to die out—as did also the greater scourge of Nazism, which had caused millions of deaths. By 25 April, partisan formations and Allied troops had freed the Po valley, where the worst massacres of all had taken place, and the entire northern region from Piedmont to the Veneto as well. On 28 April, Mussolini, disguised as a German officer, tried in vain to evade his doom, which preceded that of Hitler and the Nazi chiefs by only a few days.

The death of Mussolini recalled my father's predictions. At the beginning of the 1930s, when the Duce enjoyed the highest prestige in Italy and abroad, Father had seen him parade on horseback surrounded by the Fascist hierarchy, with his chest puffed out and his haughty air, amid the crowd's wild applause. On returning home, Father commented, "He'll wind up like Cola di Rienzo." I was struck by this reference to the fourteenth-century Roman tribune who, after having committed innumerable abuses and crimes, was put to death by a mob. Like Cola di Rienzo six centuries before, the Duce sought to escape in disguise. Both were discovered and killed, and their bodies mutilated by the same populace that had applauded them in the past and now vented its rage in an almost identical way on their corpses. Thus ended the Fascist regime, denounced *in extremis* by its most ardent supporters. In a country devastated and ravaged by war, but nonetheless more aware than it had been in 1943 when it welcomed with jubilation the news of Mussolini's fall from power, the definitive defeat of fascism was finally celebrated.

CHAPTER 13

The Return to Turin and to Research

TOWARD the middle of May 1945, because of the services I had been and was still performing as a physician, the Allies let me travel by military truck to the north, by then liberated from the Germans but still not accessible to civilians by means of regular transportation. Only Gino was allowed to come with me. We stood all the way in a truck packed with American soldiers whose gaiety and light-heartedness we could not share so anxious were we to get news and to see friends and relatives again. Those who had sought refuge in Switzerland—like our sister Nina and her family, and Gina Fubini with her husband and children—had survived but not yet returned to Piedmont. We had already been informed of the tragic end of others who had fallen into Nazi hands and disappeared into concentration camps.

It was already dark when we arrived in a seemingly deserted Milan, scarred by bombing and wearing still the drab uniform of war. There was no point in trying to find a hotel. Along with hundreds of other refugees, we settled on the steps of the Central Station's great staircase, and at dawn took the first train for Vercelli, the nearest station. On rented bicycles we then proceeded to Caserana, Mariuccia's small home town, where her father was waiting for us. Her mother had been struck down, while still young, by a fatal form of hepatitis, an endemic disease in the Vercelli region, on the very day little Emanuele was born. Gino had not told Mariuccia the news until two months after the fact, lest it increase the difficulties she was undergoing in nursing her baby.

From Caserana, we went to Asti where Mother's brother had

stayed in hiding and managed to escape capture. Upon returning to Turin, I found my friends, the majority of whom had just come back from fighting as partisans.

Two months later, we returned to our native city just as it was reacquiring the features familiar to me since childhood, no longer made ludicrous by the graffiti exalting the Duce and the glories of his new empire. But if fallen masonry had been rebuilt and the old Piazza San Carlo, which had suffered more than others from the bombs, had rediscovered the severe, monarchic appearance so dear to De Chirico, many things had changed inwardly; and it was not easy to resume the rhythms of a life so brutally interrupted by the storm of war. With the disappearance of the danger that had threatened the survival of many of us, so also disappeared the strength or blind daring that enables one to overcome life's most difficult moments. I fell into a depressed state similar to those so frequently manifested during convalescence from long and serious illness.

Most difficult of all, for me, was to recover the enthusiasm that had sustained me when I created my little laboratory à la Robinson Crusoe in Turin and later in the small refuge in the Asti region. The difficulties I had then to face prevented me from fixing my attention on the meaning of the problems I had set myself to study. A successful experiment, a perfect silver impregnation of an embryo's nervous system, were the only immediate aims of my efforts and all the more rewarding because achieved at such great expense in labor. The situation now was radically different. Professor Levi offered me my previous position as assistant. But unlike Levi who, with his inexhaustible energy, simply resumed the activity he had been engaged in for more than half a century, I did not see how the neurobiological experiments I had taken up again with Rodolfo Amprino on my return to Turin afforded any possibility of confronting—much less solving—the complex problems we had posed ourselves.

Aware of the inadequacy of my scientific training, I enrolled in biology, regretting that I lacked sufficient aptitude to enroll in the faculties of physics or mathematics. I was frustrated by the comparison of my own research with the extraordinary results obtained by geneticists from the time, in the last century, when Gregor Mendel had laid the foundation of the field, to the first decades or our own when Thomas Hunt Morgan had directed research onto the little fruit fly known as *Drosophila,* making this insect the hero of biology in the

1930s. In addition to this research, an approach that became common domain when Morgan was awarded the Nobel Prize in 1933, we in Turin were proudly following the successes of Salvador Luria—by then a professor at the University of Indiana, in Bloomington—which we knew about through articles appearing in American scientific reviews and echoed in interviews and commentaries in Italian newspapers.

It was at about this time that Renato Dulbecco and I met again in Turin, and our friendship really began. He had returned from the Russian front where he had, as a doctor, experienced the tragic lot of the Italian troops. Back in Turin he had been reappointed to the university's Institute of Pathological Anatomy as assistant lecturer, similar to my post in the Department of Anatomy of the medical school with Guiseppe Levi.

One day Renato came to see me and asked for my views on some research he was doing on lung lymphatics. From our first year in the university, I had been impressed by his marked bent for mathematics and physics and asked myself frequently why he had not enrolled in the physics department instead of medicine. Our mutual reluctance to form a friendship had prevented me from telling him of my impression. That day, for the first time, since he asked for my opinion, I felt authorized to give it. Instead of answering his question, I suggested that he should enroll in the physics department. "Afterward," I said, "you can go back to biology, and investigate much more thoroughly problems of much greater interest than the one you mentioned." I advised him to transfer from pathological to normal anatomy where his supervision of students' lab sessions would be much less time-consuming; and he consented to letting me ask Levi if he could offer him an assistant's position. Levi was willing to hear me out. I told him of Dulbecco's renewed desire to take up research in the biology of growth—though I didn't mention the fact that the reason behind the choice was essentially to have more time for studying physics. The Master was in good humor and asked me with a mischievous smile whether I had decided to take up Dulbecco's cause because, as Levi put it, I was infatuated with him. The professor's lack of psychological penetration had thrown him off the track once again. I hastily assured him that he had no need to worry on that score since nothing could be farther from the truth. Levi agreed, though still doubting the truth of the matter, and the next day invited Dulbecco for a talk and

112

TOP LEFT: Adele Montalcini
with her firstborn, Gino,
1902. ABOVE: Adamo Levi, 1905.
LEFT: Adele Montalcini and
Adamo Levi with their children,
Anna *(on the left)* and Gino,
1905.

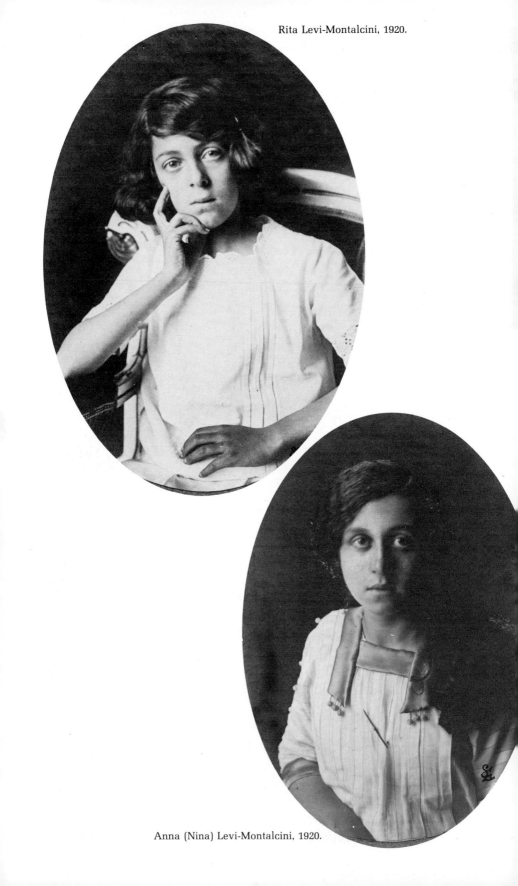

Rita Levi-Montalcini, 1920.

Anna (Nina) Levi-Montalcini, 1920.

Bronze head of Giovanna Bruatto,
sculpted by Gino after her death
in 1930.

Rita and Paola Levi-Montalcini
in the woods at Forte dei Marmi
after the promulgation of the
racial laws, 1938.

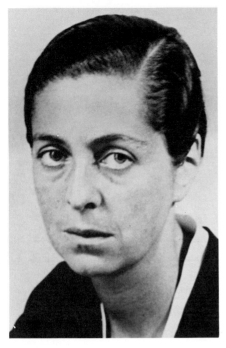

TOP LEFT: Rita Montalcini after her
return to Italy from Brussels, 1940.
ABOVE: "The Walking City," oil painting
by Paola Levi-Montalcini, 1953.
LEFT: The "Master," Giuseppe Levi,
circa 1946.

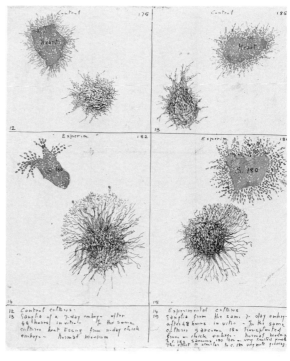

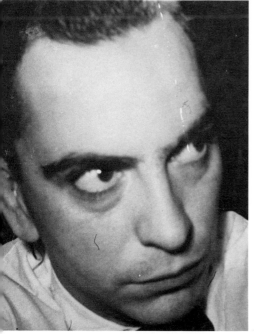

TOP LEFT: On the *Sobieski,* 1946.
ABOVE: Hand-drawn illustrations of the
in vitro effects of normal mouse
tissue and Sarcoma 180 on chick embryo
ganglia, sent to Viktor Hamburger
from Rio de Janeiro, 1952.
LEFT: "Nando," 1953.

In the laboratory at Washington Universi in St. Louis, 1960.

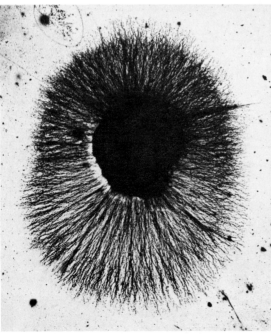

Microphotograph of a sensory ganglion explanted from a seven-day chick embryo, featured in *Science*, 10 January 1964.

Gino and Rita Levi-Montalcini, 1970.

Viktor Hamburger and Rita Levi-Montalcini, 1978.

Paola Levi-Montalcini, in front of one of her engravings
on copper, 1978. (Photograph: courtesy of Roberto Granata)

offered him the post. Renato followed my advice and enrolled in the physics department.

My doubts were resolved in an altogether unexpected quarter. On a summer day in 1946, Levi called me into his study and showed me a letter dated 8 July he had just received from Professor Viktor Hamburger, chairman of the department of zoology at Washington University in St. Louis, the author of the article that had stimulated me to repeat in my homemade laboratory his neuroembryology experiments. Having read our publication in the Belgian *Archive de Biologie*, he was curious about my conclusions regarding the mechanisms governing the effects of peripheral tissues on the development of the nervous fibers that innervate them—conclusions completely different from his own. He had written to ask Levi whether I might be allowed to spend a semester in St. Louis in order to investigate the problem further with him. While the idea attracted me, I did not want to interrupt too brusquely the biological studies I had just begun on the neuroembryological research project I was investigating with Rodolfo Amprino. I therefore decided to postpone my departure for St. Louis until the following year.

I had, not long before, written Luria asking whether he could offer Renato Dulbecco the possibility of working for a year with him at the University of Indiana. At the same time that I was deciding to accept Hamburger's invitation, I received a letter from Luria replying affirmatively to my suggestion. He subsequently offered Dulbecco a fellowship. The two of us subsequently embarked for the United States together (see chapter 15).

All that year of 1946, a common research project had allowed me to get better acquainted with Rodolfo Amprino. In the early spring, we were in the habit of eating lunch on the Anatomy Institute's top floor, overlooking Corso Massimo d'Azeglio. The trees along the avenue, just coming into bud, still allowed glimpses of the green waters of the river Po which cuts off the city from the surrounding hills. Bleaching near us were the bones of skeletons set out in the sun to dry, after removal of the muscles and soft tissue that had covered them in life. The subtle odor of decomposition from bones not yet entirely fleshless has remained in my nostrils and brings back those hours most vividly. It did not in those days interfere with eating our sandwiches nor with our conversation. I made the most of those hours to enjoy the sun, which was becoming hotter as spring gave

113

way to summer; and Rodolfo, still convalescing from a pulmonary condition that had endangered his life, drew sustenance from the healing light and enjoyed it with me. Thirty years later, we would remember those hours as among the most serene in our lives. We had both passed through our different but painful experiences of war and, in the prime of our productive capacities if not of our youth, we looked forward with joy to the future opening up before us. Neither he nor I were to be disappointed.

114

PART III

A New Life

CHAPTER 14

A New Continent

ON 19 September 1946, Renato Dulbecco and I sailed from Genoa on board a Polish ship, the *Sobieski,* I heading for St. Louis and he for Bloomington. In the course of the two-week voyage, the little *Sobieski* courageously plowed through storms which seemed would sink it at every turn. Untroubled by seasickness, I contemplated with joy the huge waves that broke on the deck, and gloried in the sun, the wind, and the salt-laden air. Forsaking the pool, where each day the other travellers spent long hours swimming, Renato and I would find shelter indoors and discuss the prospects before us. Since he had decided—if the right occasion presented itself—to settle down in the States, he was planning to have his wife and two children, whom he had left in Turin, join him as soon as possible. For myself, I considered such a choice impracticable. Mother and Paola would never have considered moving to the United States. I, therefore, thought it imperative to return at the end of the six to nine months mentioned in Hamburger's letter of invitation. These reasons of sentiment, which prevented me from considering other solutions, continued to be accompanied by tormenting doubts about the results I could hope to attain from my research. Renato, in his fervor at the prospect of studying molecular biology in the States, shared my perplexities about the future of experimental neuroembryology.

The *Sobieski* finally reached the end of its journey. When the Statue of Liberty became visible against the sky of the port of New York, loudspeakers told passengers to stay where they were and not to rush to the side nearest the statue. Two years previously, a liner repatriating several thousands of soldiers from Europe had almost capsized because of their sudden rush to the rail to cheer the statue, which to them meant not freedom so much as homecoming. Contemplating it from the distance specified by the loudspeakers, I felt as hundreds of thousands of refugees have felt, in the flight from recent as well as earlier persecutions, upon arrival in New York Harbor.

Disembarking took from six in the morning till two in the afternoon, so slow and thorough at that time was the inspection of papers and luggage before immigrants were allowed in. During the two days I spent as a guest of cousins in New Jersey, quite near New York, I payed the tribute all new arrivals feel they owe to the Empire State Building and, from its top, admired the peaks of other skyscrapers standing out like the crests of our Alps against a luminous, early October sky. Below, thronging the streets of Manhattan, bustled a huge crowd; and I was cheered to think that, in the coming weeks and months I would not be one of those minute black dots, moving about like molecules in the grip of Brownian motion.

I was taken, two days later, to St. Louis by the *Spirit of St. Louis,* a train whose name commemorated Lindbergh's historic crossing of the Atlantic. I remember it as the most luxurious train I had ever had occasion to ride on. If this memory is only partially true, the impression certainly was largely influenced by the remembrance of the cattle cars on which I had traveled during the war. And particularly of the cattle car on which had ripened the idea of repeating Hamburger's experiment, the idea that had led to my present journey to the Midwest. The liking I had had for trains since childhood, *Il Libro dei Treni (The Book of Trains)* having been my favorite reading during periods of recovery after the usual childhood illnesses, was entirely satisfied by the *Spirit of St. Louis.* If since my childhood the notion of trains and handsome puffing locomotives had been indissolubly tied up with convalescence, this train journey, undertaken in a time of convalescence from wartime experiences, was to act as a powerful restorative. Rediscovering the laws of Pavlov's conditioned reflex, I wondered why he became so famous merely for having shown that gastric juices increase in dogs at the sound of a bell that previously has signaled the imminent arrival of food. Each one of us in the course of a life goes through similar experiences hundreds of times. The train seats covered in blue velvet, the fascinating landscape beyond the windows, and a courteous hostess increased my sense of comfort and well-being, still so alien to the postwar Europe I had left.

On my arrival in St. Louis, in the somnolent hours of a hot afternoon at the beginning of autumn, Union Station appeared to me as out of a painting by the Belgian artist Paul Delvaux, in which the hands of a big tower clock mark the hours of a time that seems in reality fixed for ever. I was attracted by the crude red of the station's bricks, the tower, and a certain stately atmosphere that prevailed both in the building and in the large square in front of it. It evoked the beginning of the century when hundreds of trains a day set off from there, carrying freight and a motley crowd toward the North and the West, lured by the mirage of fortunes to be made in those still almost virgin regions of the continent. A taxi driver with a marked German accent took charge of my bags, and we set off for Washington University, a fair distance away, over on the west side of the city. He proudly pointed out the large Catholic cathedral on Lindell Boulevard which, so he informed me, had cost three million dollars and at least challenged, if it did not actually surpass, St. Peter's basilica in Rome in size and beauty. That at least was his opinion. The central gateway of Washington University, surmounted by two towers and flying the university flag, fulfilled my expectations more than the cathedral. The brick buildings of the various faculties were partly hidden by vines and separated by great lawns, which in later years disappeared to make room for more buildings and parking lots for an ever-increasing number of cars. These were still scarce in 1947, and the lawns crossed by splendid avenues served as a rendezvous for young couples in love and as a place where solitary students preparing for exams could concentrate. Young instructors also held their lessons there on those still warm autumn days, sitting on the grass with their students around them in a ring. The girls knitted while they listened, a custom that pleasantly underlined the informality of the American system of teaching but ran counter to my strong sense of equality between the sexes and to my innate dislike for this so typically feminine activity. Looking at those girls, I recalled the formality of both lectures and surroundings at the Institute of Anatomy where I had spent many of my university years. The Rebstock Building, which housed the Zoology Institute where I was heading, was made even more attractive than the others by the thick tangle of ivy crowning the doorway and the façade shaded by two great trees.

Passing inside, I found myself in front of a tall, thin young woman with a stern expression and blond hair braided behind her head. She scrutinized me and, when I told her my name, muttered something

119

unintelligible. I guessed, nevertheless, that she knew about my coming since she said that Dr. Hamburger was waiting for me in the library on the second floor. Never having seen a photograph of him, I feared I wouldn't recognize him, but he came to meet me the moment I entered the room. He smiled and nodded to me to follow him to his office across the corridor. At first, I felt somewhat at a disadvantage, partly because of the difference in our heights—he being six and one half feet to my five feet three—and the gray hair which, framing his high forehead, contrasted with his otherwise youthful appearance; but mainly perhaps because of his reputation. Sitting in front of him, however, I felt more at ease, our heights being then, if not equal, at least less uneven, and I could observe him as closely as I had been observed by the blonde downstairs (I subsequently learned she was Florence Moog, a well-known biologist). I never found out what Hamburger's first impression of me was; mine of him was certainly positive. His expression, serious and amused at the same time, with which he followed the movements of my hands while I tried to answer his questions in my broken accent, and his perfect mastery of the English language, which sounded more intelligible coming from him than from Florence Moog or the custom's officials who had questioned me on my arrival to New York, made our relationship easy from the start. In the hour-long conversation that followed I realized with joy that I had landed in the right place, and stifled the doubts that until then had been assailing me. I promised myself to make the most of being able to dedicate myself full-time, and in the most favorable surroundings, to the research I had conducted amid the disadvantages of my private laboratory. With great pleasure I accepted Hamburger's invitation to dine at his home, an attractive house in a residential district a little more than a mile from the campus. We set off on foot along one of the tree-lined avenues, and I marveled for the first time at the splendid autumnal colors of the leaves in the setting sun, which have no parallel to the dull tones of the trees in the hills around Turin during the same season. Even if habit later diminished my sense of wonder before the parade that Nature puts on before falling into winter lethargy, I still link that amazing spectacle with the recollection of my first contact with Washington University and the Midwest.

We were welcomed by Martha, Viktor's wife, also of German origin but, unlike Viktor, not Jewish. She met me with warmth and simplicity, dressed in the neat and unpretentious style of a good wife and mother. In spite of her kindness to me, I could not fail to notice a feeling of tension in the looks she gave to Viktor—tension that was to break out three years later with the violence of a hurricane, smashing the apparent calm and serenity of their family life. Martha, a woman of lively intelligence and deep culture, wanted to follow a career in her chosen field of sociology; and whether frustration at not being able to fulfill this ambition influenced her mental breakdown is impossible to say. Three years later her behavior suddenly changed. Visual and auditory hallucinations transformed whoever approached her into monsters and assassins. It was deemed necessary to commit her to one of those sinister mental institutions of the past—in which therapy consisted of the physical and psychic degradation of inmates by means of ice-water baths, strapping to beds, prohibition of books and magazines, among other deprivations. Prevalent at the time was the belief—not entirely eradicated today—that mental illness is a form of divine punishment that has been inflicted upon deviants. In the years that followed, when I visited Martha with Viktor in one of these establishments, I was struck by the dignity with which she submitted to the treatments prescribed for her and her fellow inmates. The serenity she had lacked during the happier moments of her life helped her survive the infernal circle into which she had been plunged. She died many years later, after being released from the institution and entrusted to the care of one of her two daughters.

After that first dinner at his home, Viktor accompanied me to the house of the people with whom I was to stay, a middle-aged couple living not far away, on Delmar Boulevard. Along the way he told me he had decided on the provisional arrangement of a room in a lodging house to give me the opportunity to meet a genuine American couple right away, so as to be able to get some idea of American virtues and defects. I was met by my landlady, a stocky little woman with gray hair and bright demeanor. She introduced her husband, who was in a wheelchair and suffered from aphasia. As a welcome, he mumbled something unintelligible. The wife, who could talk for both and was eager to do so, informed me that he belonged to the Lion's Club. She then rapidly brought me up to date on her political, religious, and social convictions: she was an ardent Republican, went dutifully to church every Sunday, and hated the Jews, who kept their shops open on Sunday and thus waged unfair competition with Christians. I was too tired and content with the experience of my first day in the Midwest to be upset by the differences in our opinions. I reckoned that, if the negative aspect of what Viktor had defined as the way of thought of the average American had been made apparent in such a surprisingly short time, the positive aspect would certainly be manifested soon enough. In the days, weeks, and years that followed, it was indeed. And it was one of the reasons, if not the only one, that my stay on the other side of the ocean lasted—though with long intervals—for thirty years and not for a semester as I had planned.

CHAPTER 15

Experimental Neurobiology in the First Half of the Century

What is a nerve fiber for, anyway? And what, in fact, is a nerve cell? The intriguing mechanism, by means of which it generates and conducts an impulse, has chiefly preoccupied three generations of neurophysiologists with the performance of too many and ingenious experiments. . . . Conduction of an impulse is in fact somewhat incidental to another essential functioning of the neuron, however useful as a sign that the neuron has functioned. Where does one come out, if he looks at a neuron as a secretory organ? This proposal is not new, but nobody has done much lobbying for it. To wit, the prime function of a neuron is to produce and to apply to other tissues a chemical activator.

—GEORGE BISHOP

"WHAT everyone believed yesterday, and you believe today, only kranks will believe tomorrow." With this prophecy, as he called it, Francis Crick concluded his brief yet dense discussion on the theme "of molecules and men" which deals with, among other topics, vitalism. According to this theory, which had a considerable vogue in the eighteenth and nineteenth centuries, vital phenomena cannot be adequately explained by the laws of physics or chemistry but are instead the result of a form of energy termed the "vital force." Vitalism was dealt a severe blow in 1837 by the German chemist Friederich Wholer who, with his teacher, Justus von Liebig, successfully synthesized organic molecules like uric acid from inorganic elements, thus prov-

123

ing that the synthesis of such substances is not the exclusive prerogative of living beings. Nonetheless, until not many years ago, vitalism
continued to exist in one guise or another. In this regard, Crick quoted
the opinion of the Hungarian-born, American Nobel Prize–winning
physicist Eugene Wigner that it is impossible—given data inferred
from quantum mechanics and statistically manipulated—to conceive
of systems endowed with the power of self-replication. On the basis
of such arguments, another physicist, Walter Erlasser, hypothesized
the existence of laws called "biotonic" which, though not in contrast
with the laws of physics, are not part of them either. The mysterious
phenomenon we call "consciousness" was cited as evidence of the
existence of such biotonic laws. It was this same phenomenon, Crick
noted, that caused many physicists to become interested in molecular
biology, spurred not so much by the wish to prove the validity of
vitalistic theory as by the desire to come up with irrefutable evidence
of its falsity.

Vitalism, Crick explained, is based on a series of arguments that
have increasingly lost favor. The first of these is the belief that there
exists a clear line of demarcation between the properties of animate
and of inanimate systems. The second, not substantially different
from the first and still today the subject of animated debate, is based
on the fact that it is impossible to know how life on the planet
originated. The third, the aforementioned argument sustained by
Wigner, is that it is impossible to explain the nature of consciousness
and the capacity to elaborate thought on the basis of the laws of
physics. However, Crick observed, "When facts come in the door,
vitalism flies out the window." This is what happened when the
American biochemist Wendell Stanley showed in 1937 that an organism such as the tobacco mosaic virus—which has the properties,
common to all viruses, of self-replication and of infecting other organisms—can be crystallized without losing this supposedly exclusive
property of living creatures.

On the question of how life came to exist on the planet, two theories currently contend with one another. The first, advanced by the
Russian scientist Alexander I. Oparin, is known as the "primordial
soup" theory. In the course of the billion and a half years preceding
the appearance of the first living beings (minuscule organisms known
as prokaryotes found embedded in the earliest geological strata and
not substantially different from present-day bacteria), life would
have originated from the chance interaction of organic and inorganic

substances already present in large quantities in the ocean depths. The second hypothesis—advanced by scientists such as Francis Crick, Enrico Fermi, Leo Szilard, Carl Sagan, and Fred Hoyle, to mention only some of its illustrious advocates—is known as the "guided panspermia theory." It maintains that life on our planet is the result of inseminations by living forms from other planets, not even necessarily part of our solar system. The fifteen billion years believed to have passed since the universe arose out of the Big Bang are, according to this theory, ample time for life in other parts of our own or in other galaxies to have reached higher forms than those present on our planet. This possibility, according to Crick, should not be classified among those that are the object of science fiction but should rather be considered a genuine, though somewhat premature, scientific theory. The problem of establishing the origin of life is, according to him, too important to be ignored by anyone interested in the origins of man and of consciousness. Along with the molecular biologist Gunther Stent, who like himself has been converted to neurobiology, Crick thinks that the questions of the origin of life and of consciousness do not differ substantially from the question of the transmission of traits from individuals of one generation to those of the next—a question that about only a century ago was still considered unanswerable. "In the same way that the virus known as bacteriophage provided the key to unlock the mystery of heredity," Stent writes, "the neuron, that is, the elemental unit of the nervous system, will allow us to solve the mystery of the phenomenon of consciousness."

The elucidation of a phenomenon that has fascinated humans since they first became aware of their individual consciousness, and is still wrapped in the deepest mystery, not only presents enormous difficulties we cannot know for certain will ever be overcome, but is also surrounded by an emotional aura which prevents it from being objectively and serenely investigated. Even though all inquiry into the phenomenon we call consciousness could not in the past, and still cannot, be anything but speculative and strongly influenced by the religious or atheistic convictions of the inquirer, the brain has been identified as the organ that elaborates, albeit in mysterious ways, sensations both from the outside world as well as from its own activity in the organism: thus, we are at once actors and spectators of what happens within us and around us.

In the last decades of the nineteenth century, direct, though still rudimentary, investigation of the structure and function of the brain became possible for the first time. It was the great merit of the Italian histologist Camillo Golgi to have discovered the method, still used today, of chrome-silver impregnation, which makes nerve cells stand out dramatically in the smallest detail, dark against the golden-yellow background of surrounding tissues. Using this technique, and another also based on the strong affinity of nerve cells for metallic salt impregnation, the great Spanish neurologist Santiago Ramón y Cajal succeeded in the formidable task of identifying the thousands of different nerve cell populations which—wrapped in an extremely dense interlace of nerve fibers and surrounded by an astronomical number of satellite, or glial, cells—form that organ known as the brain. So complex is it that the brain of any single human being is estimated to have roughly one hundred billion nerve cells. The entire system consists of the nerve-cell populations, the strands of fibers which stem from them and form the circuits that interconnect the nerve cells, and billions of glial cells interposed among them.

It would have been impossible to identify and reconstruct these cells had Ramón y Cajal not resorted to the strategy of analyzing the system at the early stages of embryonic development, when it largely consists of a few hundred cells and when the first neuronal circuits can be examined under the microscope. The specimen he used for this purpose, and is still today preferred above those of other species, is the chick embryo. Its nervous system, especially the cerebral component, is different from that of mammals, but the fundamental unit, or neuron, of the chick's brain—that is, the cell body and the fibers stemming from it—is identical to the basic unit of the nervous system of all vertebrates including man. The high affinity of the nerve cells of birds for silver-salt impregnation, and the mere three weeks between the time the embryo begins to develop and when it hatches, make it an ideal subject for the investigation of the architecture of the nervous system. As Ramón y Cajal first pointed out, the formation and differentiation of nerve cells—both along the cerebrospinal axis and in those formations positioned at the side of the spinal cord and brain stem known as sensory and sympathetic ganglia—take place according to laws so rigorously constant and identical in all individuals of the same species that they allow one to identify with perfect accuracy the developmental stage of any specimen under examination. On the basis of his studies, Ramón y Cajal formulated the neuronal theory which identifies nerve cells as the basic units of the nervous system, and the nerve fibers, known also as neurites or axons, as the filament extensions that connect both the cells along the cerebrospinal axis to one another and the motor or sensory cells—via the encephalic and spinal nerves—to peripheral organs. This was the first theory to see the nervous system as made up of cells—a property that had been ascribed to vegetable and animal tissues since the middle of the nineteenth century.

The histological techniques available before the discoveries of Golgi and Ramón y Cajal's did not allow for recognition of the cellular units that are the indivisible and fundamental element of nerve as of all other types of tissue. Added to the technical difficulties was the reverential aura which led to the brain's being considered as different from other organs and inaccessible to direct investigation. Ramón y Cajal's theory proclaiming the neuron the anatomic, genetic, trophic, and functional unit of the nervous system was accepted by eminent scientists such as the Swiss Wilhelm His—who advanced the theory almost at the same time as the Spanish neurologist; the Germans, M. von Lenhossek and W. Waldeyer, who coined the term *neuron;* and, in Italy, Ernesto Lugaro, Giuseppe Levi, and few others. Nevertheless, the theory met considerable opposition. Foremost among its opponents was Golgi himself. In his acceptance speech, when being awarded the Nobel Prize along with Ramón y Cajal in 1906, Golgi expressed his firm refusal of the neuronal theory and his preference for the theory that sees the entire nervous system as formed of cells united one to the other in a "diffused anatomic net," in which the fibers are materially and inseparably connected. In a bitter argument, which took on a note of personal rancor, he tried to demolish the theory of his "illustrious colleague," Ramón y Cajal, and declared the neuronal theory definitively dead. According to Golgi and to his followers, nerve impulses travel uninterruptedly through the fibrous circuits from one neuronal cell population to another like electrical current in a net of metallic wires.

Though it was clearly demonstrated in the first two decades of the century that this reticular theory, as it was called, was incorrect, I still had the opportunity to meet one of its last remaining advocates in Chicago in 1949, at the first conference on neurobiology I attended in the United States. He was the old "reticularist" Jan Boeke. His eyes, deep blue as the porcelain of his native Holland, with their somewhat fanatical expression and his proud demeanor, brought to mind the last survivors of Garibaldi's army whom as a child I had seen parading in wheelchairs or leaning on sticks, wearing their characteristic jackets and red berets, in the patriotic processions of 1918.

As Ramón y Cajal wrote, the reticular theory had disastrous consequences, delaying by at least two decades the advancement of knowledge of the structure and function of the nervous system. Of particular importance instead, and with extraordinary implication for research into the function of the nervous system, was his intuition that not only is every cell an independent unit, but that the ends of their fibers—which through an optical microscope appear to be closely linked to the cell body or to other components of the neurons with which they communicate—are not in fact materially connected with these cells, but rather are separated from them by spaces so small as to mislead even the most careful user of an optical microscope. It was the English physiologist C. S. Sherrington (who played the same pioneering role in the field of neurophysiology as Ramón y Cajal played in neuroanatomy) who called these spaces *synapses,* a name still used today. With the invention and application to biological studies of the electron microscope, whose power of magnification is thousands of times greater than that of the most powerful optical instrument, it became possible to see the synapses, which vary in size but never exceed 300 Angstroms.* One of the main merits of the neuronal theory over the reticular, in fact, was its assumption of spatial discontinuities long before the technical means were available to validate the hypothesis.

One might ask oneself how a theory that viewed the nervous system as an indivisible whole, a theory contradicted by everything already known about the functional properties of the system's various sectors, could not only have been given credence by persons of known merit but also last as long as it did. A reply has been suggested by the English biologist M. Abercrombie:

> The meaning of this controversy between neuronists and reticularists is to be sought not only in the importance of elucidating the basic mechanisms which govern the functioning of the nervous system, though its importance be obvious. It represented a frontal clashing between those who believe that the mystery of an organism is lost when it becomes the

*The Angstrom is a unit of length equal to 10^{-7} millimeters.

object of analysis and dissection into its various components, and those who believe that the atomistic approach is justified and that it is possible to reconstruct from partial results the global functioning of the whole.

In the case of the nervous system what played in favor of the reticularists, as I have suggested, was the mystery and religious awe with which people approached the organ that elaborates cognitive faculties believed to be the unique privilege of individuals of our species and a direct emanation and proof of its divine essence.

After the definitive acceptance of the neuronal theory, the central issue became whether the different areas of the cerebral cortex perform different functions or whether the latter are the result of the activity of the whole cortex. In the middle of the last century, the French neurophysiologist Paul Broca had already proved that circumscribed lesions in the third frontal circumvolution of the left cerebral hemisphere provoke a pathological process known as aphasia—namely, the impossibility of expressing oneself by means of language. A few years later, a young German neurologist, Carl Wernicke, had localized the faculty of understanding spoken language—as opposed to speech—in a circumvolution of the temporal lobe in the same hemisphere. But the problem of localization emerged again in neurological research at the turn of the century, and was tackled in the following thirty years by such eminent scientists as the American Karl Lashley in the then-nascent field of experimental psychology and the Austrian-born American Paul Weiss in that of experimental neurobiology.

In a long series of studies on the capacity of adult rats to recognize and solve specific problems before and after their brains have undergone both limited and extended excisions of specific cortical areas, Lashley came to the conclusion that the capacity to learn and memorize the outcome of past experience is not localizable in well-defined areas of the cortex but, rather, is the result of its global activity. The excision of small areas did not affect such capacities, which were instead lost after extensive excision. Lashley's conclusions concurred with the Gestalt theory, whose major exponent was the German Wolfgang Köhler. According to this theory then much in fashion, perceptive impressions are the result of the spatial and temporal summation of the stimulation of individual sensorial elements; from their ensemble, and not from each one separately, the individual forms a global idea of the world and of the events that send different messages to his or her brain.

The interest and assent that greeted Lashley's findings were in turn supported by the results of the neuroembryological studies carried out by Paul Weiss. Weiss applied techniques from experimental embryology, a discipline born toward the end of the last century and in full development at the beginning of the present one. First in his native Germany and then in the United States, Weiss sought to explain the mechanisms involved in the distribution of nerve fibers among peripheral organs. There were two conflicting theories: one, that the growth of the fibers and their connection with the cells of peripheral tissues and organs is the result of a pre-established program unmodifiable by the environment; the other, that such growth and connections take place in a way that is not predetermined, but shaped by the receiving organs. Using cinematographic means to observe the contractions of supplementary limb muscles grafted onto amphibian larvae and fully grown specimens, Weiss found that the muscles responded—according to their differing physico-chemical characteristics—only to those nerve impulses that their constitution allows them to receive and to contract to. His theory was based on the study of the motility of grafted limbs, which turned out to be synchronous with and identical to that of homologous limbs in the host. On the basis of his results, Weiss reached the conclusion that the growth and ramification of nerve fibers are not predetermined, but are subject to the influence of the innervated organ. His theory in agreement with Lashley's, attributed to the cells, to the neuronal circuits, and to nerve tissues an almost unlimited capacity of adapting to environmental stimuli both in the central and in the peripheral nervous systems.

It was in this antideterministic climate that a pupil of both Lashley and Weiss, Roger Sperry, then at the start of his career, vigorously challenged the theories of his two teachers. "The rebellious graduate student"—as Viktor Hamburger defined him many years later when he conferred on him the Ralph Gerard Prize for the Neurosciences (one in the uninterrupted series he received for his universally acknowledged contributions to psychology and to the study of the distribution of nerve fibers and of their functions to peripheral nerves and intracerebral fiber bundles)—conducted systematic studies with extraordinary rigor, using simple methodologies. He demonstrated the nonvalidity of the principle of equipotentiality and the nonadaptability of nerve fibers and cells to the most varied manipulations of the experimenter, such as the cross-switching of the nerves in muscles with opposite functions. He also observed the regeneration of the optic nerve of lower vertebrates whose ocular bulbs had been severed.

Sperry studied both the behavior of rats whose nerves had been first sectioned and then reconnected to muscles with opposite functions, such as the extensors and flexors in the lower limbs; and the visual function in fish and amphibians that had undergone surgical severing of the optic nerve and rotation of the bulb (in lower vertebrates, as opposed to higher ones from birds all the way up to mammals, nerves and fiber bundles in the central nervous system are endowed with the capacity to regenerate and resume their functions). His conclusions, based on behavioral and biological studies resulted in a categorical refutation of Lashley's and Weiss' theory that fibers grow and regenerate in an entirely haphazard manner. On the contrary, the results of Sperry's experiments reaffirmed the principle advanced by Ramón y Cajal that nerve fibers are governed, during their growth and in the course of regenerative processes, by a rigidly predetermined genetic program which fixes the direction of connections formed with other nerve cells in the centers or with tissues and organs in the periphery. Results obtained with both lower and higher vertebrates showed irrefutably that the formation of synaptic connections does not occur in a diffuse and chaotic fashion but rather takes place according to a principle of high selectivity. Sperry suggested that the latter might in turn be the result of the complementary specificities of nerve ends and nerve or other types of cells with which synaptic connections are established.

My first encounter with Sperry and Weiss occurred on a cold day in March 1949, in Chicago, where the latter had organized a conference on the topic, very ambitious for those days, of "genetic neurobiology." It was the first conference ever held on the different aspects of the newly established field of neurobiology, and the participants, among whom were Viktor Hamburger and me, were about twenty in all. In recent years neurobiology conferences take place simultaneously in various localities and gather together hundreds of speakers and thousands of participants. Sperry was then in the midst of his research work. I was struck by his resemblance to Gary Cooper, the hero of films of that era. Throughout the entire conference, which lasted three days, Sperry remained silent just like the movie hero with whom I had associated him, thus blatantly expressing his complete lack of interest in the other speakers' papers. "Gary Cooper is dreaming about his prairies in the Far West," I murmured to my neighbor, Alberto Stefanelli, a young neurobiologist who had come from Rome. Thirty-two years later, the Nobel Prize Sperry received for his studies on neuronal specificity and on the functions of the two cerebral hemispheres, fully recompensed Sperry for the difficulties he had encountered at the beginning of his career. But in 1949, his tenacious opposition to the theories of his direct superior, Weiss, made it difficult for him to pursue the experiments he had begun, and his future as a research scientist was uncertain. I had the impression that, though remaining silent, he followed with interest my report on the processes of migration of thousands of cells in the chick embryo's nervous system at early stages of its development.

While early neurobiology was struggling in the quicksands of vague theories propped up by experimental data of dubious validity, neurophysiology, which has as its object the analysis of the nervous system's functions, was making important advances and attracting an increasing number of researchers thanks to the discovery and utilization of the cathode-ray oscilloscope. This instrument, when used in combination with a photographic camera, allows one to register the image painted on a screen by the spontaneous or exogenously induced electrical activity of nerve cells, both single and in groups. The highly amplified image of the registered signal enabled researchers to record the nervous activity of particular centers under specific, chosen conditions. The cathode-ray oscilloscope's application to the study of the electric discharges of nerve impulses into peripheral nerves—which contract spontaneously, either by way of reflex or induced stimuli from the innervated muscles—allowed them to be first recorded by the English physiologist Edgar Adrian in 1925. This was a date of historic importance for neurophysiology: from then on, thanks to this powerful instrument, it became possible to investigate the mechanisms that give origin to nervous impulses and the way they travel along nerve fibers. There was still great debate about whether the nervous impulse, once it has arrived at the terminal of the fiber in close apposition to the recipient cell, is transmitted to the latter by means of electrical activity or by a chemical substance, or

132

"neurotransmitter," released from the nerve endings. The argument between the advocates of the two theories, known as the "battle between the soup and the spark," ended many years later when it was recognized that, in the majority of cases, the transmission of the impulse from a fiber to the activated (or inhibited) cell occurs by way of the liberation of a neurotransmitter of a chemical nature, whereas in a limited number of cases, and especially in the nervous system of invertebrates, it occurs by way of electricity.

What contributions have these discoveries made to the solving of the fundamental problems of neurobiology, whose object is the study not only of the structure and function of the basic unit, the neuron, but, in particular, of the set of billions of cells making up that immensely complex organ known as the nervous system and the brain? Interesting in this regard are certain remarks quoted in the epigraph to this chapter, of a pioneer in these studies, one of the most outstanding neurophysiologists of the first half of the century, George Bishop, whom I had the good fortune to know well since he was for many years and until his death director of the department of neurophysiology of Washington University Medical School: Everyone agrees with Bishop that the generation and the conduction of the nervous impulse and the recording of the action potentials of groups of cells—and then of individual cells, which became possible after John Eccles introduced the splendid technique of impaling nerve cells—cannot, by themselves, elucidate the function. It was, however, the possibility of recording the potentials of cells situated in peripheral locations, and along the different stations that convey the nervous impulse from the peripheral organs to the cerebral cortex, that allowed for the charting of the routes that nervous impulses follow from the moment they are generated in the peripheral organs to the instant they are received by the cortical centers.

Studies on a variety of important topics—such as, to mention only a few, the nature of neurotransmitters and neuromodulators, the ways in which they activate and/or inhibit recipient cells, the differing properties of the brain's two hemispheres—have opened to neurobiological investigation vast horizons, which to the participants gathered around the table in that March of 1949 would have seemed pure science fiction. I have been fortunate to witness the transformation of a sector of biology previously considered among the most static into one of the most active. The thirty scholars who participated in the conference on genetic neurology, including the old "reticularist," have today become thousands of young investigators in all branches of science, from neurology to molecular biology, from immunology to physics, all of whom are engaged in the search for the thread of Ariadne which may lead into those cerebral recesses where vitalism's last stronghold is waiting to be stormed.

CHAPTER 16

Explorations: The Midwest and the Developing Nervous System

IN 1947, at the time of my arrival, St. Louis was a growing city of about two million inhabitants who were very proud of their past. From the time of its foundation in 1763–64 when Pierre Laclède and August Chouteau, two French fur traders, picked a spot on the west bank of the Mississippi River not far south of its confluence with the Missouri as a staging post for the fur traffic, the original core had expanded and changed to become the city named after Louis IX, the crusader king of France. In its early days, the inhabitants were mostly French, coming from New Orleans and Canada as well. Their descendants live in St. Louis today, and those original French settlers are remembered in the wide boulevards that bear their names: Laclède, De Balivière, Chouteau, Ladue, Lafayette, Bonhomme, and so on.

Among St. Louis's many glories were its having joined the Union side in the Civil War and offering refuge to blacks fleeing from the South.

Toward the end of the century, the city became one of the most flourishing in the United States. At the time of my arrival, the memory was still vivid of the famous world's fair of 1904, which had brought to St. Louis twenty million visitors and where were displayed, for the first time, the silent film, the wireless telephone, and the dish-washing machine. No less important as objects of attraction were the famous St. Louis blues, jazz, and that curious discovery and immediate success—the ice cream cone.

Also toward the end of the century, immigrants were coming to St.

Louis from Germany, Ireland, and Italy. The latter, mainly from Piedmont, were later joined by a larger wave of emigration from the south of Italy. The Italians, who settled in an area in midtown called the "Hill," were one of the few ethnic groups who did not intermarry but stayed on the whole united through the customs, language, and dialects they had brought from their places of origin. Characteristic of St. Louis's Little Italy were the shops that sold typical Italian foods, an attraction for those outside the neighborhood, and certain features of the streets and houses which made one feel one had landed in a southern Italian town. On one of the last Sundays in August, the residents of the Hill let themselves go in dancing, singing, devouring ice cream, and all things Italian. There was a groundswell of pride of a nationalistic-Fascist kind. For them, Mussolini was not only not dead but still enjoyed a prestige equal to that of Guglielmo Marconi who was considered the symbol of Italian genius. A friend of mine, a well-known physicist of Jewish origin who had been living in the United States for several years, had been asked on one of these feast days to say something to celebrate the discoverer of radio waves. Though he had spoken in praise of the great physicist, he had also—with little feeling for the circumstances and his surroundings—referred to a few of Marconi's negative aspects, among which his well-known support for fascism. My friend's words were received in icy silence, and later someone commented that it had been in extremely bad taste to let the greatest Italian genius be commemorated by a Jew. Along with fascism, there survived a streak of anti-Semitism which had found fertile ground in the descendants of immigrants from Italy, who identified their patriotic feeling with the Fascist ideology. In my rare visits to the famous shops on the Hill, I avoided any reference to politics and especially to fascism.

Shortly after my arrival, I had a clash with the then Italian consul who, in a gathering of common acquaintances, had manifested his hatred for the partisans who had fought in the Resistance, and his regrets at the fall of the regime. It was my first and last encounter with him. I became good friends, instead, with an Italian couple, Paolo and Silvia Rava—a friendship founded on our common background and cultural interests. Though a few years younger than I, they "adopted" me into their family and not a Sunday or holiday went by that I wasn't a guest at their house; I also regularly turned to Paolo in his capacity as an able lawyer for advice in regard to the many problems that beset new arrivals to the United States. Our friendship continues to this day—as does my affection for their daughter Lucy, a woman of extraordinary sensibility gifted with the talent for seeing only the positive aspects of life and the better qualities of the persons she meets.

Every now and again, overcoming my natural disinclination for whatever was not directly connected with the work I had set myself—in what I believed would be but a short stay in the Midwest—I would go downtown, where buildings famous for the severe lines of their nineteenth-century architecture—such as the post office—lined the main avenues. The latter were flanked by narrow alleys where the poorest inhabitants, mostly blacks, lived in slum housing. This part of the city extended as far as the west bank of the Mississippi, with its muddy stream upon which intense river traffic had once plied. The only remaining relic of this past was a boat, with the glorious name of *Admiral,* on which people took Sunday outings. On these trips, I would sit apart from the mainly teen-age crowd and their dancing and singing, and enjoy from a deckchair the burning rays of the summer sun, an ice cream cone, and the panorama of small islands in the stream flowing between banks of thin forests through which could be seen small houses and vegetable patches that reflected the poverty of their dwellers.

On Sundays, I alternated between trips on the *Admiral,* excursions organized by people from the department to the picturesque Ozarks—mostly picnics on small boats along the rivers and lakes of that area—and trips to relatively nearby Bloomington, Indiana, to see my friends from my student days, Salvador Luria and Renato Dulbecco. It was not nostalgia that drove me to these latter train or bus journeys, but the desire to talk with them about my projects and to get their opinion, which I valued highly. I retained, in fact, an unconfessed envy for geneticists and microbiologists, whose work I believed was far more interesting and promising than that of neuroembryologists.

Luria, by then considered a foremost microbiologist, was the right person to consult. The first time I went to Bloomington, he noticed I was troubled by something. He invited me into his study and encouraged me to "tell him everything," using the Piedmontese dialect which, though we both spoke it atrociously, gave our conversation the familiar tone I needed. He spoke and gestured in a way we both knew belonged to the famous professor Nino Valobra, the family doctor of the Jewish set and of the upper middle class and aristocracy of Turin. Though Valobra was old and rather fat, with red hair and the small pointed beard typical of the illustrious physicians of his time, and Luria was black-haired, young, and quite thin, they had the same sharp eyes. Luria produced a marvelous imitation of him, assuming the same bored, meditative air, closing his eyes as if lost in deep thought, while asking the patient not to digress but, rather, to state the reasons that had led to the consultation. I told "all" to Luria. I liked Washington University and Viktor Hamburger, whom he also knew, if only slightly. (Viktor had, in fact, asked him for references about me after reading my article, and it had been Luria's flattering recommendation that had prompted Viktor to invite me to St. Louis.) I did not hide from "Doctor" Luria my distressing doubts about the validity of our approach to the problem of the differentiation of the nervous system. I believe his doubts were much greater than mine, but he saw no way out of the difficulty. He excluded the possibility of my successfully taking up microbiology, unprepared as I was in the field, and encouraged me, without enthusiasm, to continue my research.

He ended the consultation and said that he wanted me to meet the most brilliant of his students, who had recently become part of his team. The person in question was James Watson, then not yet twenty. He had the appearance of an adolescent, his brow still shaded with blond hair which was to give way to premature baldness. His absorbed and dreamy look, his slenderness, and his absentminded way of moving reminded me of a famous Picasso Harlequin of the Blue Period. He took no interest in me whatsoever and left immediately with a hasty goodbye. Our brief encounters over the following years, when the unknown adolescent became the famous Watson, were characterized by his absolute indifference toward me—an attitude I saw as part of his well-known anti-feminism. I was never troubled by it.

After Watson, I met the two most celebrated scientists in the University of Indiana's department of biology, colleagues and friends of Luria's—Tracy Sonneborn and Hermann Muller. The latter had, the previous year, been awarded the Nobel Prize for his discovery of the mutation effects of ion radiation, but both were, to me, equally eminent.

Explorations: The Midwest and the Nervous System

With Sonneborn, I had merely a brief and courteous exchange. Our interests differed, and I confess that at the time I was completely ignorant of his classic studies on *Paramecium aurelia,* the unicellular organism also known as protozoa. In the autumn of 1947, it was firmly believed that he would receive a Nobel Prize. Diminished interest in the problems that he was investigating deprived him of this recognition, though not of his reputation as a major scientist.

My meeting with Muller, the subject of whose research fascinated me, lasted longer. Contrary to my expectations, he showed—possibly out of pure politeness—interest in my research, and I was astonished and flattered by the attentive way he listened to what I had to say about it. Both he and Sonneborn were typical examples of what Tullio Terni used to define as people who are "all nucleus," that is, lacking those cytoplasmatic fringes that in those days were believed to be of only secondary importance. Sonneborn and Muller looked and behaved like the image I had had since my student days, with my fanatical devotion to science, of what real scientists were like: namely, people different from and superior to all other living beings. During the long course of my life, I was radically to change this point of view and reexamine my scale of human values.

I used to return to St. Louis from these rare encounters with Science with a capital *S* in an unaltered, depressed state of mind. In spite of my pessimism with respect to the work I was doing at the time, the nervous system still exercised such a strong fascination upon me that I found it hard even to envisage abandoning the studies I had begun. Furthermore, I was anxious to reinvestigate with Viktor the problem that had prompted him to invite me: the question, that is, of the effects of amputation on the development of the nervous centers in charge of the innervation of the excised limbs.

About a week after my arrival, I had left the boardinghouse where I had first stayed, with its ardent Republican keeper who hated Jews, and found lodging in a beautiful room as the boarder of an old lady in her eighties who lived about half an hour's walk from Washington University. Mrs. Elizabeth Gray seemed the living image of a past era in her grace, in the delicate features of her face, in the elegance and wit of her conversation. She was a faithful Democrat who had shared my joy at the news of Truman's election, did not frequent church, and derided her bigoted acquaintances. She took a liking to me, as I to her, from our very first meeting. The only thing that troubled her was my habit of coming back home late at night, walking from Washington University to De Balivière along the avenue flanking Forest Park. Even if crime in those days was a much less widespread phenomenon than it became later, the park still had a reputation as a dangerous place in the evening and night hours. But I could not afford a car at the time, and public transport was so infrequent that I preferred the long nighttime walks. Often a soft caress would startle me in the middle of the night. I would see her frail aged figure naked by my bedside. "Sleep well, my darling," she would murmur, "I came to make sure that you were home." In the months that followed, her mental lucidity progressively waned. Eventually her son decided to have her committed to an old-folks' home against her wishes. Whenever I went to visit her, she assured me that she was in perfect health; she was confident that she would soon be allowed to return to her home, and asked me if I would be willing to continue to live with her. I assured her that was my intention and left her happy at the thought of still spending many more years together.

My doubts about the future of my research, which had led me to seek Luria and Dulbecco's advice and been the cause of many sleepless nights, were not to last long. From one day to the next and in an entirely unforeseeable way, I regained the faith in and the enthusiasm for the possibilities of experimental neuroembryology that I thought irretrievably lost.

It happened in the course of one afternoon in late autumn of 1947, while I was somewhat haphazardly examining, under the microscope, the latest series of silver-salt–impregnated chick-embryo sections. Their coloring had come out perfectly: the nerve cells that had begun to differentiate in the cerebral vesicles and spinal cord—the embryos had been fixed between the third and the seventh days of incubation—stood out in their every detail with a deep brown-blackish hue on the golden yellow background of the nonimpregnated cord tissue formed of satellite cells and as yet undifferentiated nerve cells. My attention was drawn to the spinal cord which revealed, even at first glance, a surprising variety of scenarios not only in its different segments but—in embryos at the same stage of development and fixed just a day or a few hours apart—also within the same segment. The thoracic and sacral segments offered a spectacle not unlike that of the maneuvers of large armies on a battlefield. Thousands of cells in the thoracic segment, and a smaller number in the sacral one, were proceeding in long columns from the ventro-lateral to the dorso-medial sectors of the spinal cord. The migration began on the fourth day of incubation and ended on the sixth, when the cells reached their destination at the sides of the central canal. The identical modalities with which, in all of the cases I examined, the various phases of the process took place, revealed the genetic programming at work and, at the same time, the striking similarity between these processes and those that—in respect to the migration of birds, insects, locusts, termites, and ants—we view as revealing the workings of "instinct."

In the case of the migrations of living organisms, one can fall back on this vague term which explains everything, and therefore explains nothing. In the case of cells, unable as we are to use the term *instinct* since we do not attribute to them even hypothetical powers of decision, one speaks of the activation of a genetic program, not essentially different from programs that programmers draw up for computers. The thin dividing line between the realization of an instinct and of a program lies merely in the assumption that the former is a property of multicellular organisms in which we recognize the presence of a grouping, rudimentary as it may be, of nerve cells in charge of an individual's behavior; whereas the latter is common to living entities such as unicellular organisms or cells, and to inanimate objects such as computers as well. That day, however, as I peered through the microscope, nerve cells were acquiring a personality not usually attributed to them.

At the cervical level, I was witness to the disappearance of the membrane marking the boundaries of the nucleus, the retraction of the nerve fiber, and a decrease in the volume of the cell bodies. In immediately successive stages, the same cells acquired an appearance known, in histological terminology as "picnotic," or indicating the establishment of a process of irreversible degeneration. The impression was that of a battlefield covered with corpses. In embryos fixed at the stage immediately afterward—that is, ten to twenty-four hours later—the ventro-medial sector was invaded by macrophages, or cells able to ingest and destroy bacteria and the detritus of other, degenerated cells. The image that suggested itself to me was of corpses being removed from a battlefield by special crews trained and equipped for the purpose. A few hours later, all traces of cellular detritus had vanished; the column of cells still differentiating was

141

vastly reduced in diameter, and an inspection of the fibers of the cells that comprised it revealed that they were innervating the muscles of the same segment of the trunk.

The scenarios I observed through the eyepieces impressed me not only because of their dramatic and dynamic character, but also, and even more, because they showed that the nervous system employed a strategy completely different from what had until then been attributed to it. It had been commonly believed, in fact, that quantitative and qualitative differences among cell populations in the central nervous system are the result of the different proliferative activities of the various segments of the cerebral-spinal axis. It now appeared clearly evident that the nervous system makes use of a different tactic and provides for the different functions carried out by nerve cells in one of several ways: by means of drastic eliminations within the ranks of excess, unnecessary cells—as was the case in the cervical spinal cord; by means of the active migration of cell populations in charge of different functions, such as somatic and visceral activities; or by way of the separation of cell populations with the same function but destined to innervate different peripheral tissues.

The reconstruction of these many processes—which took place in my mind in the span of an hour—filled me with joy. It struck me that the discovery of great migratory and degenerative processes affecting nerve cell populations at the early stages of their development might offer a tenuous yet valid path to follow into the fascinating and uncharted labyrinth of the nervous system. The doubts about the future of experimental neuroembryology, which had prompted me to seek Luria's advice, had been based on the conviction that it would have been impossible to unveil the mechanisms underlying neurogenesis with the theoretical and technical means then at our disposal. The absence of valid criteria to follow in such a venture was even more evident in the light of comparisons between this field of study and those of genetics and virology in which—thanks to excellent models and sound, verified principles—always newer and greater heights were being reached. Not only did the nervous system not lend itself to experimental analysis, but its enormous complexity seemed to discourage all attempts to espy whatever might be taking place in it, "behind the scenes" as it were. The startling realization that nerve cell populations were subject to quotas and to the elimination of excess numbers in their ranks, as well as to migrations that went hand in hand with functional differentiation, showed that there were ontogenetic processes at work in the nervous system which were not as inaccessible to investigation as I had previously imagined.

In that intoxicated state, I knocked on Viktor's office door and asked him to follow me back to my lab bench. Scattered next to the microscope were the hundreds of slides containing the sections I had just examined. I rapidly sketched the images I had observed on a piece of paper, and told him my interpretation of the processes as I had mentally reconstructed them. Viktor listened attentively, amused by my enthusiasm, and agreed that they were observations of extreme interest, which provided the key for the study of differentiative processes in the nervous system up till then practically ignored. His enthusiasm, rising to the same pitch as my own and making me all the happier in light of his habitually more calm and cautious attitude, was confirmation of what since the start had struck me as a sort of revelation. When Viktor left the room, I put one of my favorite records on the record player I had installed in the laboratory: a Bach cantata.

If it had taken but an hour to reconstruct and interpret the meaning of the images I had seen at the microscope, a much longer time was to be spent in the rigorous reconstruction of the processes and their description in an article that saw the light only a year later. Though in the years that followed, I was to taste the joy of discoveries of far greater import, the revelations of that day stayed permanently inscribed in my memory as marking not only the end of the long period of doubt and lack of faith in my research, but also the sealing of a lifelong alliance between me and the nervous system, an alliance I have never broken or regretted keeping.

CHAPTER 17

The Nerve Growth Factor: The Opening of the Saga

THE DISCOVERY of the migratory and degenerative processes that affect nerve cells revealed to me the existence of vast unexplored areas of the developing nervous system and reconciled me with neuroembryology, which now appeared to be a field much more open to experimental investigation than I had previously thought. At the time, I wrote Mother and Paola: "I smell a stimulating fragrance of truffles in the air. I trust my sense of smell and think I have found the scent-trail of a large specimen. This time [I had referred earlier in the letter to previous discoveries that had produced no advantages to my career) I am determined to put it to personal use."

One morning in January 1950, the university campus was covered by a thick blanket of snow and a freezing wind, typical of the Midwest, had caused the temperature to plummet to $-20°$ C. In a state of joyous anticipation, I was busy surveying a series of recently sectioned embryos in the warm, $+20°$ C. atmosphere of the laboratory, when Viktor appeared at the door. He told me he had received a letter from an ex-student of his, Elmer Bueker, who had, a couple of years before, moved to New York where he had continued to investigate the effects of peripheral tissues on the differentiation of the nerve cells innervating them. In his letter, Bueker described a phenomenon that puzzled Viktor, who in turn wanted to know what I thought about it. The data, which were reported and also illustrated in a brief 1948 article of Bueker's accompanying the letter, seemed extraordinarily interesting to me, much more even than they appeared to the author.

Bueker's experiments represented an ingenious variation of those

144

originally performed by Viktor and repeated by me in my small laboratory in Turin—and which still today are an object of study for both of us. He had analyzed the same problem: namely, the regulatory mechanisms governing the development and differentiation of motor and sensory nerve cells in chick embryos. To this end, instead of grafting additional limbs, he had grafted fragments of neoplastic tissues endowed with great proliferative activity, such as chicken Rous sarcoma and mouse tumors. It was already known at the time, on the basis of experiments performed many decades before, that all sorts of mammalian tumors, from rodent to human, cling and grow when transplanted on the vesicled membrane that envelops chick embryos—since organisms at the embryonic stage do not yet possess the immunizing properties that cause tissues from donors of other species to be rejected. Taking advantage of this fact Bueker had assayed the propensity of randomly selected tumors to develop after being grafted onto the body of a chick embryo. One of the tumors he used, the murine (or mouse) sarcoma known as S. 180, took hold and formed a small neoplastic mass between the embryo's posterior limb bud and its abdominal cavity. Bueker noticed that, in embryos fixed on the eighth day of incubation, the nerve fibers stemming from nearby sensory ganglia had ramified into the mass of tumoral cells. The author concluded that the tumor had provided a more ample terrain than the limb for the fibers to grow in. According to his interpretation, S. 180's properties and rapid proliferation produced conditions favorable to the fibers' growth—an effect that, in turn, caused a greater number of nerve cells in the ganglia to differentiate, resulting in the latter's increased volume. In his letter, Bueker asked Viktor for an opinion on these experiments, which he had performed two years previously and then abandoned.

In my euphoric and "anticipatory" (as I defined it in the letter I was writing home at the time) frame of mind, the effect Bueker described struck me like a message in cipher whose meaning it was up to us to discover. Viktor agreed with me that we should interrupt the experiments we had in progress, which consisted in the grafting of limbs and other normal tissues, and repeat those Bueker had described. Flattered by the interest we took in his work, he wrote that he was pleased we were going to repeat it. Ten days later, we received from Jackson Memorial Institute (an institution that specializes in raising mice with malignant tumors, which are kept growing by means of

145

serial transplants) a boxful of small albino carriers of S. 180. In many, the tumor had grown almost as large as the mouse, but the animals did not seem encumbered by their mass, and appeared full of vitality as they inspected their new cages with their minuscule, ruby-colored eyes. I put to sleep the ones with the largest tumors, cut open their abdominal walls, and transferred the neoplastic masses into glass capsules. I then took fragments of the tumors the size of chick-embryo limb buds and grafted them onto the sides of three-day-old embryos.

Subsequent inspection of embryos fixed one to two weeks later revealed, in the majority, a well-circumscribed nodular mass, interlaced with vessels, which protruded from the abdominal cavity next to—and partially invading—the embryonic kidney on the same side. As I examined at the microscope the tumor-carrying embryos, which had been colored using the silver-salt–impregnation method and serially sectioned, an extraordinary spectacle presented itself to my eyes. The mass of tumoral cells, which differed from those of the embryos in being larger and more intense yellow and in having a different growth pattern, were from all sides penetrated by bundles of brown-blackish nerve fibers. These fiber bundles passed between the cells like rivulets of water flowing steadily over a bed of stones. A closer inspection of the individual fibers revealed another singular fact: in no case had they established any connection with the cells, as is the rule when fibers innervate normal embryonic or adult tissues. Tracing the origin of the bundles, I was surprised to discover that they had originated for the most part in the sympathetic ganglia. A smaller percentage of fibers originated from the sensory ganglia adjacent to the tumor. The volume of these ganglia, and even more of the sympathetic ones, appeared vastly increased with respect to the volume of the ganglia that innervated limbs, blood vessels, and organs on the other side of the embryo. These observations, which were confirmed in dozens of cases, fueled the impression I had had upon reading Bueker's letter that the effect caused by the grafting of S. 180 tumor fragments was indeed very different from that provoked by the transplanting of supranumerary limbs, and contained a "ciphered message" as it were. Since the start of these first experiments, Viktor had been in Cambridge, Massachusetts, meeting some commitments at M.I.T.; but I kept him up to date with long weekly reports. I had the feeling—a feeling that with the passing weeks became ever more certain—that I had come upon a phenomenon without precedent in the rich case history of experimental embryology.

The Nerve Growth Factor: The Opening of the Saga

Then, on a day in the autumn of 1950, the "block" was suddenly removed, and what had thus far retained the veiled character of an impression assumed the distinct connotations of absolute certainty. The effects of graftings of S. 180 were indeed *radically different* from those of limb buds and other embryonal tissues.

I have never been able to clarify to myself whether this revelation, so like the sudden parting of a curtain, was the result of a great accumulation of data in my unconscious or whether the mice in a second shipment from Jackson Memorial Institute were carrying tumors that, as chance would have it, actually differed from those of the first batch. The order forms accompanying the two shipments stated that the first batch of mice were carriers of S. 180 while the second group had been grafted with a tumor known as S. 37. Though classified as sarcomas, a term commonly used to indicate tumors of connective origin, both had been derived from mouse mammary carcinomas that, in being serially transplanted, had lost the original structure characteristic of the mammary gland and of the carcinomas affecting it, and had begun to grow in a disordered and tumultuous fashion—though whether from cells of epithelial or connective origin it was impossible to say.

In the course of that morning in the autumn of 1950, I had been examining at the microscope the effects of a series of graftings of S. 180 and S. 37 in embryos fixed after eleven days' incubation—eight days, that is, after the transplanting of the tumors. The scenario that presented itself to my eyes was so extraordinary that I thought I might be hallucinating. Not only were the tumoral masses being invaded by a thick network of nerve fibers stemming primarily, as in the cases I have already described, from the sympathetic ganglia of the paravertebral chains and prevertebral ganglial complexes, but so also were the as-yet noninnervated organs of the embryo itself. In the mesonophros—the organ that in the early phases of the embryonic life of vertebrates precedes and fulfills the same function of the metanephoros, or permanent kidney—the fibrous bundles had spread among the tubules without establishing connection with the cells coating them. A chaotic and random distribution of fibers was also visible among the other organs, such as the primordial sexual organs, spleen, thyroid, tymos, and liver.

These extraordinary and atypical effects bore no relation to those resulting from the grafting of supranumerary limbs; another effect suggested to me how this phenomenon might be correctly interpreted.

147

I saw that the sympathetic fibers not only had invaded the viscera of the embryos but had also perforated the walls of blood vessels and entered inside the small and large veins, while avoiding the arteries. Having bored their way through the veins' sheathing, they had assumed the form of large, protruding, fibrous bundles (known as "neuromas") inside the vessels themselves. The smaller veins were, as a result, completely obstructed; in the larger ones, the tangles of fibers appeared as round formations afloat inside the vessel. This observation indicated that the tumors had released a humoral, or fluid, factor able both to accelerate differentiative processes in sympathetic and, to a lesser degree, in sensory cells; and to cause the precocious and excessive production, as well as the quantitatively and qualitatively abnormal distribution, of nerve fibers. The penetration of the nerve fibers into the veins, furthermore, suggested to me that this still-unknown humoral substance might also be exerting a neurotropic effect, or what is also known as a chemotactic directing force, one that causes nerve fibers to grow in a particular direction. Unlike arteries, the veins were draining away from the embryonic tissues the waste products released by the neoplastic cells. Among these, I guessed, was undoubtedly also the humoral growth factor that the cells produced. This hypothesis would explain this most atypical finding of sympathetic fibers gaining access inside the veins.

The validity of the latter hypothesis was confirmed on the basis of in-vivo and in-vitro experiments performed many years later. Now— with the hindsight of the nearly forty years gone by since those moments of keenest excitement—it appears evident that the new field of research that was opening before my eyes was, in reality, much vaster than I could possibly have imagined.

In the days and weeks that followed, I examined dozens of other embryos grafted with these and other tumors. Whereas some of the latter grew even faster and more abundantly than S. 180 and S. 37, none exerted even the slightest stimulating effect on the fibers of sympathetic, sensory, or other nerve cells. Every time I was due to examine embryos with graftings of either of the two original tumors, I was overcome by fear that they would have ceased to exert their magical effect. In those moments I thought of the trepidation the people of Naples experience each year on 3 September when, in the course of the feast honoring San Gennaro, Naples's patron saint, a crowd of devotees packs the church and awaits the miracle of the liquefaction of the saint's blood. In the ceremony, unchanged since time immemorial, the officiating priest solemnly lifts the precious vial containing the blood of the saint and holds it before the crowd. Usually, the blood is slow to liquefy, to the congregation's great anxiety. It is believed, in fact, that if the miracle fails to occur, the city and its inhabitants will suffer terrible plagues or other signs of divine wrath. If the delay continues, the descendants of the San Gennaro family, who have always enjoyed the honor of sitting in the first pews, hold themselves responsible to their fellow citizens for the missing miracle, and rage against their great ancestor. They express themselves with both anger and familiarity: "Do us this grace, yellow old man, scapegrace saint!" The saint responds to the threatening requests of his relatives: without fail, the blood is seen slowly trickling along the inside of the vial, as the crowd exults. Whenever, while looking in the microscope, I failed immediately to notice the ramification of fibers among the tubules of the mesonephroi, I was tempted to curse, but then exulted, too, upon witnessing, once again, the miracle of the humoral factor.

It was on one of those intense mornings, while I was busy examining sections at the microscope, that I was surprised by the familiar voice of Professor Levi thundering in the hallway. I knew he was in the States, but not that he would be coming to St. Louis. I was all the more glad of his visit in the expectation that he, more than anyone, would appreciate my discovery. Briefing him on my experiments, I begged him to look at the preparations through the microscope. Levi listened to me, looked through the eyepieces, and shook his leonine mane, which despite his age (he was then over eighty) was still red and barely sprinkled with a few gray hairs. Finally, he gave vent to his rage, in a sudden fit that for me evoked the long-gone days of my internship when his anger had filled me with terror. Had I in so short a time forgotten everything he had taught me? How could I be so blind? Even a beginner would have immediately understood that those were not nerve—but, rather, connective—fibers. Did I not remember that fibers of connective origin sometimes become black when treated with silver salts? I was going to ruin my reputation and, indirectly, his own if I persisted in my interpretation and communicated my "discovery" to a periodical. Knowing from experience how useless it would be to try to convince him, I shuddered at the prospect of spending all the days of his visit to St. Louis locking horns over the question of the fibers' identity. Suddenly, I had a good idea, and asked whether he had ever seen the Grand Canyon—a hypocritical question, as I knew perfectly well that this was his first trip to the United States (also, as it turned out, his last). In a few minutes, he had agreed to take immediate advantage of St. Louis's relative proximity to Arizona to see that stupendous panorama. Three days later, he flew back from Phoenix, in an excellent mood and completely covered with snow, then falling thick in St. Louis.

I had taken advantage of his absence to illustrate—using a device that projected images appearing through the microscope on pieces of paper placed alongside it—the course of the nerve fibers stemming from the sympathetic ganglia into the viscera and veins of the embryos. The old lion glanced distractedly at my reconstructions. "May be," he grumbled, "but I don't believe it. Nor will you ever succeed in convincing me that a tumor can change the way nerve fibers grow." I did not insist. We spent the rest of his last stay at the museum and in my favorite bookstore. Wishing to buy me a souvenir of his visit to St. Louis, he purchased a huge history of Germany because—as he gruffly put it—he had become aware that I was entirely ignorant in matters both historical and geographic. I confess to never having had the time to read that tome. He also gave me a beautiful volume of reproductions of paintings by El Greco, whose works in the museum we had both admired. Indeed, they excited him to a degree astonishing to me who knew the limitations of his artistic culture (matched only by mine in history and geography).

As soon as I had taken my old Master to the airport, I conceived of an experiment that would convince not only Levi, but all other potential objectors as well, of the validity of my hypothesis on the humoral character of the factor released by the two tumors. I decided to take advantage of the fact that chick embryos are wrapped in an extra-embryonic membrane (known as the corio-allantoic membrane) which is crossed by a thick network of blood vessels belonging to the embryo itself. The function of the membrane is to protect the embryo while permitting gaseous exchange, to serve as receptacle (in the allantoic vesicle) for waste products, and to enable the embryo to receive nutriment, from the thick net of capillaries which cover it, by conveying nutriment-carrying blood from the egg yolk to the embryo's umbilical vessels. I grafted small fragments of S. 180 or S. 37 close to the small and larger vessels on the corio-allantoic membrane. Each day, through a small opening in the calcareous eggshell, which I kept covered with an easily removable piece of adhe-

The Nerve Growth Factor: The Opening of the Saga

sive tape, I watched through the stereomicroscope the development of the transplanted fragments. In the majority of cases, the transplants had taken hold, appearing like little nodules surrounded by a dense vascular network. If my hypothesis was correct, the small neoplastic masses would release into the circulation the humoral factor to which I attributed the effects I had observed after the intra-embryonic transplanting of the tumors. In this case, however, I could exclude the possibility that the effect was the result of the conveyance of the factor along nervous pathways, since the corioallantoic membranes are not innervated and communicate with embryonic tissues only by way of the circulation of blood.

After ten days of incubation, I removed the embryos, impregnated with silver salts, and examined them at the microscope. In all cases, I witnessed with immense gratification that the same effect observed after the graftings on the embryos themselves had, in fact, taken place: the sensory ganglia, and the sympathetic ones even more so, were much larger than normal in volume; and sympathetic fibers had invaded the organs, some even penetrating all the way into the veins of the host embryo. These results demonstrated irrefutably that the two tumors had their effect even without any direct contact between them and the embryos, and confirmed my hypothesis that the tumoral agent endowed with the property of stimulating the development of receptive nerve cells and of their fibers was being transmitted in a way characteristic only of a humoral substance.

I wrote about my progress to Renato Dulbecco, who promptly replied: "Your news is most exciting. I read your letter with such avidity as one reads about the occurrence of a sensational event. And sensational it truly is." That was 6 January 1951. Despite my joy at his reaction, however, I was well aware that the identification of the factor, which in my letters home at the time I referred to as "the nerve-growth promoting agent," would be, if not insurmountable, certainly very difficult indeed—especially given my lack of training in, and familiarity with, biochemical techniques.

I dedicated all of my time in the following months to in-depth analysis of sympathetic nerve fiber distribution in embryos carrying intra- and extra-embryonal graftings of the tumors. I presented the results of this work, in a symposium held at the New York Academy of Sciences, in a conference that bore the not-very-inspiring title "The Chick Embryo in Biological Research." My turn to speak came at the end of an afternoon session, in an atmosphere that had been wilted by the four speakers preceding me. Despite the torpor pervading the hall, I was congratulated by the most important personalities present—such as Paul Weiss, who defined my results the "most exciting discovery of the year." I was not, however, in the best frame of mind to enjoy the praise I received, having been overcome during my talk by intense abdominal pains due to a kidney colic. Upon returning to St. Louis, I was admitted to hospital for an urgent operation.

Two weeks later, I was back in the lab attempting to reproduce the tumors' effect by injecting their extracts into embryos at early stages of development. Persistently negative results led me to resort to other techniques. In Turin, under the guidance of Levi and of his assistant and co-worker Hertha Meyer, I had learned the technique of in-vitro culture and successfully cultured cells belonging to connective, muscular, and nervous cell lines (see pp. 59–60). Levi's studies on the growth of nerve fibers from sensory ganglia explanted in semi-solid media under varying experimental conditions can even today be classified as among the most rigorous ever performed. It occurred to me that such a technique might provide me with a valid method for providing further confirmation of the humoral character of the factor and for identifying its chemical nature.

I had been exchanging letters with Hertha Meyer since 1939, when she had moved to Rio de Janeiro and set up an in-vitro-culture unit at the local Institute of Biophysics directed by Carlos Chagas. Now I wrote to ask her whether I could carry out my research in their laboratory. This solution seemed not only faster and easier than trying to introduce such a technique to our laboratory in St. Louis, where no one had any experience with it, but also had the added attraction of affording me an opportunity to visit Brazil, which I had been wanting to do for some time. Hertha spoke to Professor Chagas, who said he would be happy to have me as a guest there for three or four months. Viktor, who had approved my plan from the start, even managed to obtain a grant from the Rockefeller Foundation to cover my traveling expenses. Thus, at the end of the summer of 1952, I implanted some cells of S. 180 and S. 37 into a couple of little mice, and with them left for Italy, where I spent a few days with my family before boarding another plane headed for Brazil.

CHAPTER 18

The Fibrillar Halo
and Carnival in Rio

I LANDED at Rio de Janeiro airport on a stormy September afternoon in 1952. I was not alone: two little white mice were spying, with their pinpoint eyes, on whatever they could glimpse through the holes in the top of the small cardboard box that I had fitted into my overcoat pocket. Every now and then they took little bites out of the remains of the apple that had satisfied their hunger and quenched their thirst during the long journey from Rome. The discomfort they suffered from their confined quarters was compensated by their feeling safe inside the little box.

We were welcomed at the airport by Hertha Meyer, known as Doña Hertha in Rio at the time. In her capacity as an expert in in-vitro culture—a technique that had then been in use in biology for only a few decades—and as the assistant of the famous German biologist Emil Fischer, she had been invited by Professor Levi to move to Turin when Hitler came to power. There she had spent a happy and very productive period of her life, enjoying the great esteem and friendship of Levi, whose close collaborator she had become. When racial restrictions were instituted in Italy as well, she decided to accept the invitation of Professor Carlos Chagas, director of the Institute of Biophysics of the University of Rio de Janeiro, to organize an in-vitro culture unit in his laboratory. Almost half a century later, the unit she set up is still wholly efficient, and Doña Hertha continues to take an active part in research.

From the airport she took me and my two little traveling companions to Copacabana where I found lodging in the apartment of friends of hers. The following day, I had a first meeting with Carlos Chagas

153

at the Institute of Biophysics, which at the time was only a few hundred yards away from the Praia Vermehla, a small crescent-shaped beach under Sugar Loaf Mountain. Chagas was well known in the field of neurophysiology as a pioneer in the study of the bioenergetic processes that enable various species of tropical and subtropical fishes to generate electrical currents of considerable intensity. I therefore expected to find myself in front of a scientist of a certain age, satisfied with his successes and prestige. I was pleasantly surprised to see a young man come forward to greet me, who shook my hand very cordially and showed keen interest in my research project. His welcome and the extraordinary beauty of the view from the window of the laboratory filled me with glee. The following day I began the series of experiments that determined the direction of all my future research. They opened as well an entirely new sector of study in developmental biology by revealing the capacity of certain protein molecules—known as growth factors and synthesized by neoplastic and normal cells—to stimulate proliferative and differentiative processes in cells of various origin.

Having never kept a diary, I was very pleased when, in June 1980, Viktor Hamburger sent me a large envelope containing all the letters—carefully preserved for so many years—that he had received from me during the period I spent in Rio, from September 1952 to the end of January 1953. Along with my letters, and the India ink drawings with which I had illustrated the results of my experiments, was a letter of his containing one of his many dear and flattering comments: "These are most interesting historical documents and they show beautifully how real research works, ups and downs, despair and triumph. . . . In a way I am sad to part with the letters but they are more important for you, and they would probably get lost when I am not around."

I have reread the letters: fifty pages of onionskin airmail paper rendered yellow by time and covered with the compact handwriting my family had come to designate "chicken tracks" it was so hard to decipher. In reading them, I have relived one of the most intense periods of my life in which moments of enthusiasm and despair alternated with the regularity of a biological cycle.

Upon returning home the first day, I found my landlord's basset hound waiting for me by my bedroom door. He showed neither curiosity nor interest in my regard, and I was surprised by the assurance with which, having given me the cold shoulder, he followed me into the room. I hadn't long to wait to find out what attracted him: lying flat between the folds of a pink muslin curtain, I saw an enormous cockroach similar to, but considerably larger than, those known as *Periplanetae Americanae* which by the millions infest the South and the Midwest of the United States. "My" cockroach, or at least the one that had chosen my room as its home, belonged to the family of *Leucophaee maderae*. With incredible dexterity—all the more amazing since his little legs weren't made for speed—the basset hound leaped upon the *Leucophaee*. Within seconds, the satisfied expression on his muzzle and the sound of his sharp little teeth crushing the wings and armor of his victim informed me of his success. From then on it was I who every evening summoned him to inspect the room, accepting with a mixture of gratitude and disgust his services, almost always crowned by success.

The following morning, as instructed by Hertha, I climbed onto the footboard of a *bonde*—a means of transport that at the time consisted of cars not unlike the streetcars of the Turin of my childhood but without doors and windowpanes, these unnecessary in Rio's mild climate. I struggled in the crowd and, finding it impossible to reach the interior of one of the carriages, became a part of one of the human bunches that hung out from the platforms—all the while admiring the agility of the ticket sellers who leaped like acrobats from one footboard to another of the moving vehicle in order to check that all the passengers had their tickets.

In my letters to Viktor during the first months, I reported a complete lack of success. Fragments of neoplastic tissues removed from one of the two little mice and grown in vitro in proximity of the sensory and sympathetic ganglia of chick embryos not only had not—as I had hoped and expected—stimulated the development of nerve fibers, they had completely inhibited the luxuriant outgrowth of satellite cells and fibroblasts which in ganglia cultured alone, or combined in vitro with tissues from the same embryo, migrate away and settle in a vast area around the explants. Though not encouraging, these results seemed to me to be in any case less disappointing than if I had observed an absolute identity between the effects of normal and cancerous tissues.

The failure of fibroblasts and other nonneuronal cells to migrate away from the ganglia cultured in vitro in proximity of small aggre-

155

gates of S. 180 or S. 37 cells might have been explained by the presence of some toxic factor that masked the activity that in vivo caused the outgrowth of the sensory and sympathetic fibers from their respective ganglia. I had only one card left to play and rested all my hopes upon it: to assay the effect of the tumors that had been previously transplanted and had taken in chick embryos. The hypothesis that tumors so transplanted possess different properties from those grown in their natural, mouse-tissue environment was prompted by the results of some experiments I had conducted in St. Louis. There I had observed that the transplanted neoplastic cells differed morphologically from their nontransplanted counterparts: at every new passage from one embryo to another, the ability of the transplanted cells to stimulate the development and precocious differentiation of sensory and sympathetic ganglial cells increased. Was this the result of the selection of cellular clones that exhibited the property of synthesizing and releasing the mysterious growth factor? Even though I possessed no proof in favor of the hypothesis, in my secret heart of hearts I was certain that the tumors that had been transplanted into embryos would in fact stimulate fiber growth.

Results were far above my fairest expectation. The letter I wrote to Viktor immediately after examining the cultures through the stereomicroscope oozes enthusiasm from its every line. From the ganglia grown in proximity to the previously transplanted tumors, there stemmed an extraordinarily dense halo-shaped outgrowth of nerve fibers. In my euphoric letter, I described the halo effect and illustrated it with ink drawings in the margins: "The fibers on the side closest to the tumor are so thick and numerous that they seem to belong to a tissue whose separate components are indistinguishable from one another. . . . More interesting still," I continued after an interval, "is the fact that a similar though not quite as clamorous effect results from the addition to the culture medium of an extract of chick embryos carrying the tumors." This finding prefigured those that we were to obtain after my return to St. Louis and that were to lead to the identification of the growth factor.

The results I had described in that letter of 2 November were confirmed by the wide-scale experiments performed in the following weeks. Enchanted by the beauty of the halos, I never tired of repeating the experiment, and documented it with drawings in my numerous letters. In ganglia cultured for longer periods of time—that is, two

The Fibrillar Halo and Carnival in Rio

or three days instead of a mere twenty-four hours—not only were the fibers on the side nearest the tumor more dense than those on the opposite side; but it appeared evident that their specific orientation as well was in the direction of the neoplastic tissue, a fact I interpreted as evidence of a neurotropic effect. That one of the properties of the factor released by neoplastic cells is the exercise of a tropic, or directional, effect on the ganglia's nervous fibers was proved unassailably, however, by us and other investigators only twenty-five years later.

I was not only delighted at the sudden revelation of this extraordinary phenomenon but put also in a state of no mean perplexity. What proof was there that this in-vitro effect, so different from that induced by the grafting of S. 180 or S. 37 in chick embryos, was provoked by the release on the part of the neoplastic cells of the same factor at work in the developing embryo? Totally ignorant of the nature of the agent, I could only resort to two types of indirect proof: to check whether grafts of tumors that did not induce hypertrophic and hyperplastic growth in the sensory and sympathetic ganglia of chick embryos, induced the formation of fibrillar halos in vitro, and to check whether other tissues, in addition to S. 180 and S. 37, caused the halo to grow though deprived of the capacity to promote a similar outgrowth in developing embryos. The results of the first type of experiment confirmed my hopes: mouse adenocarcinomas, rat Rous sarcoma, and other tumors that took and grew wildly in chick embryos, but did not affect the number of fibers emerging from or on the volume of their host's ganglia, also did not cause fibrillar halos to grow around ganglia in vitro. At that point, in order to circumstantiate my evidence even further, I decided to try to culture fragments of normal mouse tissues, heart tissue and subcutaneous tissues found in proximity to the ganglia in vitro.

In a letter dated 20 November, I define the results of the latter experiments as "the most severe blow to my enthusiasm that I could ever have suffered." Though somewhat less pronounced, the in-vitro effect on ganglial fiber outgrowth of the normal mouse tissues was substantially similar to that of the S. 180 and S. 37 transplants. Just like the tumors, they inhibited the migration of fibroblasts and satellite cells away from the ganglia. Was the effect of the tumors, therefore, nothing else than a somewhat more pronounced, general "mouse effect"? Since neoplastic cells replicate faster than normal ones, could the difference be merely quantitative and, as such, entirely unrelated to in-vivo effects? I was assailed by the fear that if the nature of the agent released by the tumors—whose existence, given its imposing effects, was beyond doubt—could not be identified by means of in-vitro assays, then the phenomenon would fall into that "district of Hades most feared by scientists," as Michael Bishop once described it—"i.e., that ignominious corner called neglect."

Now, with the hindsight and greater experience of the many years since then, I wonder why finding that normal mouse tissues released the same factor as neoplastic ones evoked in me such a strong rejection reaction. An explanation may be found in the fact that in the 1950s scientists were still far from imagining that there existed specific growth factors—later identified as protein molecules—which could be synthesized and released by some normal cell lines and selectively picked up by others. The effect observed upon transplantation of mouse sarcomas appeared not only completely atypical but also probably explainable as the result of the actions of a virus that transformed normal cells into cancerous ones. In favor of this hypothesis stood the anomalous, enormous increase in the size of the ganglial masses and the atypical distribution of the nerve fibers they produced. Nothing of the sort had ever been observed after the transplanting of limb buds or organ primordia belonging to donors of the same or of different species, as the vast number of case histories compiled by experimental neuroembryologists amply documented.

After suffering the brunt of the initial shock at these results, in a partially unconscious way I began to apply what Alexander Luria, the Russian neuropsychologist, has called "the law of disregard of negative information . . . facts that fit into a preconceived hypothesis attract attention, are singled out and remembered. Facts that are contrary to it are disregarded, treated as exception, and forgotten." Thus, what I had labeled the "mouse effect" was a message I was not really capable of receiving, since I could not help thinking that it diminished—to the extent of annulling—the significance of the induction of the fibrillar halo by S. 180 and S. 37. All my efforts in the following weeks were directed toward convincing both myself and Viktor—as I, in fact, managed to do—that the difference in the effects observed was also qualitative—a conclusion even today still far from certain.

All the letters I wrote between November and the end of December, however, attest to the fact that my efforts to emphasize the differences rather than the similarities between the tumor and the mouse effect did meet with success. By Christmastime, I had fully reacquired the conviction—which later showed itself to have been well motivated—that the effects the two tumors induced in vitro were the result of the same agent that caused the precocious differentiation and increase in the overall volume of the ganglia in embryos. I wrote to Viktor a letter that sounds like a peal of victory: "The tumor effect exists!" In it I wrote of my decision to put aside for the time being the mouse effect, as "an unpleasant and complicated finding" which, however, did not in any way lessen the validity of the tumor effect. A posteriori I feel that this declared refusal of mine fully to acknowledge the implications of the effect of normal murine, or mouse, tissues proved to be advantageous. It in fact permitted me to return to St. Louis in the most favorable state of mind in which to continue investigating the factor. Some years later when, in a way both unexpected and fortunate, the mouse effect was brought back to our attention, we were finally in a position to evaluate it and to understand its meaning.

Having regained my peace of mind, I decided to dedicate the last month of my stay in Brazil to exploring Rio and its surroundings, a pleasure I had previously denied myself, so taken up was I with experiments and the ecstatic contemplation of halos. Ever since the first days after my arrival, I had been surprised and enchanted by the sense of cheer that was everywhere in the air. In the *bonde,* mingling among workers and white-collar employees were people of all ages who had the beach as their destination instead of factories and offices, people wearing beach clogs and swimming suits. A gaiety pervaded all parts of the city: in stores I was offered a *cafesino* upon entering, before there could be any talk of buying and selling. By the time I had started to allow myself a few hours of relaxation a day, Carnival was just about to begin, and the streets resounded with the sounds of singing and the rhythmic beating of drums by the troops of barefooted children to announce the nearing of the great feast.

I left town before the latter reached its most explosive phase, but joyfully joined in with the crowd who once every year pay tribute to Iemanjà, the goddess of the sea, by means of rites on the beach. Tens, perhaps even hundreds, of thousands of, for the most part, black men and women dressed in white tunics invaded Copacabana beach at sunset bearing bunches of flowers, the most varied objects, live offerings. There they lit candles, and at midnight cast everything into the ocean in homage to the goddess and in gratitude for her favors. Thousands of the inhabitants of the shanty-town *favelas,* who had come down from their shacks nestled on the hills, danced and sang along with members of the privileged classes. No mass demonstration could have more clamorously negated the racist ideologies that a few years earlier had transformed a large part of Europe into an immense prison camp. I had recently come from Florida where there existed at the time a most rigid apartheid of whites and blacks. Here all were instead a harmonious part of the same compact crowd whence songs of joy in all languages and dialects rose toward the sky. During that night in Rio, spent next to the ocean lit by the thousands of tiny flames of the torches hurled into it by the crowd, I, too, lit a candle in my mind in honor of that marvelous crucible of different races.

CHAPTER 19

Stan and the NGF

I DID NOT return directly to St. Louis. In the course of my last two weeks in Rio, I had become ever more convinced that the fibrillar halo which grew around sensory and sympathetic ganglia in vitro within a few hours of culture in a semi-solid medium in the proximity of fragments of mouse sarcomas 180 and 37, would provide us with the philosopher's stone for explaining the nature and action of the mysterious factor released by the tumors. Before setting myself to this task—a task that turned out to be far more difficult and significant than I imagined at the time—I decided to celebrate my discovery by taking a trip to Peru and Ecuador. There, amid the gigantic ruins of Macchu Picchu, I saw llamas, and Indios, who, with their handsome, olive-colored faces and sweet and melancholy expressions, negotiated in hushed tones the price of goods in the marketplaces of towns wedged into the narrow valleys of the Ecuadorian mountains. These two weeks proved a most salutary break from the orgy of halos that had coincided with the first measures of the Carnival in Rio.

Viktor was waiting for me at St. Louis airport in January 1953. I had left behind the blazing sun of South America and returned to the Midwest's freezing wind, which I didn't like but nonetheless suited my nature better than tropical climates. From Rio, I had replied to a letter from Viktor in which he informed me that he had offered to a young biochemist, Stanley Cohen, the possibility of joining our team: "From the way you describe him he seems the right person to tackle the difficult problem of identifying the factor released by mouse sarcomas." It didn't take long for me to realize how well founded my hopes were.

Stanley—or Stan, as he has been known to us from the start—had gotten a doctorate in biochemistry on the East Coast and then moved to Colorado, attracted by the earthworms that proliferate by the thousands in the humus of the Rocky Mountain forests. Having accomplished the task (I do not know whether he had set it to himself or whether it had been suggested to him by someone else), of elucidating the mechanisms governing the excretion of urea in the nematodes, he had accepted a postdoctoral fellowship that Washington University's well-known physico-chemist Martin Kamen had offered him. "I accept," Kamen had told him, "everyone who has the courage to come work with me." Stan passed the test with flying colors; and, just as his fellowship term was about to end, Viktor, encouraged by the good news I was sending from Rio, offered him a position as research associate. We met for the first time in his laboratory, on the ground floor of the institute. From that winter day in 1953 till the summer of 1959, when Stan went to work at Vanderbilt University in Tennessee, we met many times each day in that room in front of a desk fitted in between a window, in the shade of a great tree, and lab benches covered with vials and test tubes. It was in that room of Stan's, and in mine on the second floor where he often came to visit me, that we witnessed with trepidation and disbelief the sequence of events each turn of which revealed new properties of that mysterious character who, having made an appearance in Rio, received identification papers in 1954, to become known as Nerve Growth Factor, or NGF for short. Viktor, too, often took part in the meetings, even though not participating directly in the research work, and followed with enthusiasm equal to our own the always more surprising developments that emerged from our experiments.

I had been immediately struck by Stan's absorbed expression, total disregard for appearances—as evinced by his motley attire—and modesty: "If I manage to solve a problem," he told me, "it's only because I've really plugged away at it. I have to work hard, very hard, to find the solution." As he said this, he closed his eyes and lips tight, to represent the kind of effort it took him. He never mentioned his competence and extraordinary intuition which always guided him with infallible precision in the right direction. While he had an inadequate opinion of himself, he was, on the other hand, always ready to recognize intelligence in others. Of the people whom he called "bright guy" or "smart fellow," only very few have achieved fame, whereas Stan is universally recognized as a major exponent of the new field of study of Specific Growth Factors, which in those years was just being born.

Stan and the NGF

I saw Stan again recently. The thirty-four years that have passed since our first encounter have improved rather than weighed upon his physique. The glimmer that shines today as it did then behind half-closed eyelids is more mischievous, the smile more cunning, and the way of dressing, thanks to the care of his young wife, no longer the fruit of neglect. The prestigious honors of this last decade have not in the least altered his manner of acting toward others; he remains today as he has always been, entirely devoid of vanity and self-satisfaction in regard to the successes he has attained. But back in 1953, Stan was farther from conceiving that he might one day attain eminence than he was from the possibility of putting on a pair of track shoes and trying to win a running contest. He would arrive in the morning with a pipe in his mouth, limping slightly because he had had polio as a child, after traveling the short distance that separated his prefabricated cottage—one of the many scattered on top of the low hills of the campus—from Rebstock Hall, where our institute was. He was followed by Smog, the sweetest and most mongrel dog I ever saw. Smog used to lie down at Stan's feet when he sat at his desk, and kept a loving eye on him, or slept when he fidgeted with test tubes or relaxed playing the flute.

I have often asked myself what lucky star caused our paths to cross. He, too, profited from our collaboration. If I, in fact, knew nothing of biochemistry, Stan, when he joined us, had but vague notions about the nervous system. The complementarity of our competences gave us good reason to rejoice instead of causing us inferiority complexes. "Rita," Stan said one day, "you and I are good, but together we are wonderful." Coming from a person as modest and reserved as he was, this declaration surprised and flattered me. It was in this atmosphere, charged with enthusiasm and anxious expectation for the results, which kept streaming forth in a steady flow from the ground and second floors, that I lived the six most intense and productive years of my life.

As soon as I returned to St. Louis, I set up an in-vitro culture unit in my lab and confirmed the fact that only extracts of the tumors first transplanted into and then excised from chick embryos induced the formation of a fibrillar halo around sensory ganglia in vitro. The tumors that grew in the form of little nodules in chick embryos, however, contained but very small amounts of the halo-producing factor. Thus, in order that Stan might identify it, I had to provide him, for each assay, with neoplastic tissue extracted from dozens of embryos: a very tiring task which I undertook because it was the only way we could reach our goal. The latter was, in fact, reached—or so we believed—after one year of intense work: the tumoral factor was identified as a nucleoprotein, a macromolecule formed by complexes of nucleic acids and proteins (as, for example, viruses).

In the course of a conversation with the biochemist Arthur Kornberg, Stan expressed his suspicion that the nucleic component could be a contaminant, and Arthur suggested he treat the fraction with snake venom. The latter contains, among other enzymes, the enzyme phosphodiesterase, which breaks down nucleic acids. Stan grabbed the ball on the rebound. Among the fractions that I assayed in vitro the following day, there was one containing snake venom. Having not been told which of the fractions had been specially treated, I was completely stunned by the stupendous halo radiating from the ganglia. I called Stan in without telling him what I had seen. He looked through the microscope's eyepieces, lifted his head, cleaned his glasses which had fogged up, and looked again. "Rita," he murmured, "I'm afraid we've just used up all the good luck we're entitled to. From now on, we can only count on ourselves."

Events were to prove him wrong. What he called good luck had as yet only partially revealed itself, though that morning's revelation left us both amazed. The halos, which extended out in rays from the ganglia cultured in a medium containing S. 180 extract and a small quantity of venom, were of a density and symmetry without precedent. Two hypotheses presented themselves to us as most plausible: either the venom had neutralized an inhibitor in the tumor extract and caused a tremendous increase in the tumoral NGF's activity; or the venom itself contained a factor endowed with the same property of stimulating the growth of sensory and sympathetic nervous fibers. Ten hours later we were in a state of bliss as we observed the latter hypothesis confirmed. A minimal quantity of the venom fraction added to the culture medium produced, even in the absence of tumoral extract, the same dense halo that had appeared in the presence of both tumor and venom. From this new, unhoped-for source of NGF, Stan purified a fraction that, in terms of relative weight, was three thousand times richer in NGF than tumoral extracts. Having ascertained the protein nature of the component with NGF activity, Stan proceeded to determine the molecular weight and other physicochemical characteristics of the NGF extracted from the venom, while

I analyzed the effects of the same fraction injected in chick embryos. The latter experiments removed any doubt once and for all about the validity of in-vitro assays as tests for NGF activity. Injections in four-to-eight-day-old chick embryos of NGF isolated from snake venom had the identical effect as fragments of S. 180 and S. 37. The high concentration of NGF present in the venom allowed us to demonstrate what I had been unable to show using S. 180 and S. 37 extracts, where the factor was present in amounts too small to have the same effect as that of actively growing neoplastic cells. At the same time, two hypotheses had to be ruled out: that the synthesis and releasing of NGF was a prerogative of neoplastic cells; and that the factor was viral in nature.

The accidental discovery that two malignant mouse tumors and the venom of snakes share the unique property of promoting the growth of fibers in nerve cells of the same matrix—that is, the neural crest—made it seem very unlikely that neoplastic tissues and a fluid of such philogenetically disparate origin could be the only entities possessing such a property. It was, thus, necessary to extend the analysis to other tissues and substances secreted or excreted by different glands. Our research was greatly facilitated by our being able to resort to the in-vitro assay to test hundreds of samples and thus obtain, in the space of a few hours, data that either demonstrated or ruled out that NGF was being synthesized and released. Then Stan's intuition indicated a shortcut to us. The two submandibular salivary glands of mice are distinctly dimorphic: they are far more voluminous in males than in females and synthesize, in addition to other enzymes that intervene during digestive functions, a toxin that the mouse, when it bites, instills into its aggressor or victim. Significantly greater amounts of the toxin are synthesized by male mice, which are also more aggressive than female ones. The homology between the defensive and the offensive functions of the snake's venom glands and these mouse glands suggested to Stan that we begin our analyses by testing the latter for NGF. The extract was added in progressive dilutions to the culture medium of chick embryo ganglia. The following morning we marveled, with a mixture of happiness and incredulity, at results far surpassing our greatest expectations: fibrillar halos had grown around the ganglia cultured in a medium in which a couple of adult male mouse glands had been diluted 10^4 times. The equivalent of the extract of one gland in 50 liters of saline solution!

This amazing result also explained away the problem of the mysterious "mouse effect" that I had described to Viktor in my letters from Rio six years before and which had so troubled me. All mouse tissues are rich in NGF—though not nearly as much as the salivary glands are—which is released in ways that in vitro tests allow for the ready detection of.

The very high concentration of NGF in the glands revealed by the first assay and then confirmed by quantitative analyses performed on the extract, and the possibility of exploiting such a rich and easily available source of the factor, offered undreamed-of opportunities for analyzing the nature of a phenomenon that did not fit in with existing proposed schemes on the nature and the mechanisms of action of the factors regulating differentiative processes in developing organisms and in cell lines grown in vitro. NGF, in fact, was unlike hormones because of its unusual and heterogeneous sources, its ability to stimulate two types of nerve cell selectively, and the degree of response it elicited in the cells. Among the many problems that appeared to be of prime importance, three concerned the following questions: the elucidation of the nature of salivary NGF; the analysis of its spectrum of action among developing organisms and completely differentiated ones; and, finally, the function, if any, of such large quantities of the factor as were produced by the mouse glands. The first question was clearly for Stan to tackle; the second, for me to try to solve. As far as the third was concerned, since the very start its solution appeared so difficult that we decided to set it aside for the time being, resigning ourselves to accepting the NGF molecule as one of the many offerings that Nature makes without explaining their meaning. This turned out to be a wise decision. It is a well-known though often neglected rule that many apparently unsolvable problems, at one point or another unexpectedly find their solution. In the case of the submandillary glands, the latter presented itself only twenty-five years later.

I think Rebstock Hall never saw research proceed with such fervor and enthusiasm as it did from the winter of 1958 till the summer of 1959. Stan identified the salivary factor as being a protein molecule characterized—to the extent that could be educed from its chemical composition—by a molecular weight and structure similar to those of the NGF present in snake venom. I analyzed the effects of repeated daily injections of salivary NGF on newborn rodents. These experiments, which could not have been performed using NGF extracted from tumors or snake venom because of the scarcity of the former and the high price of the latter, revealed the full extent of salivary NGF's effect: that is, to increase the size of sympathetic ganglia in mice and newborn rats tenfold after only a few days' treatment, and to cause an imposing increase in the innervation of the internal organs and cutaneous tissues; an effect that, in adult specimens, was less pronounced though still observable.

In December 1958, Viktor told us news that for me sounded like the tolling of a funeral bell: budget restrictions prevented him from being able to offer Stan a permanent position in the already fully staffed department. I saw coming to an end the most productive period of my life and also the most picturesque in the saga of NGF, when Stan with his magical intuition and flute played the part of the wizard, charming snakes at will and getting the miraculous fluid to flow forth from the minuscule mouths of mice.

Before actually leaving, however, Stan gave two further demonstrations of how far his talent for "thinking hard" could take him. In the course of the experiments with snake venom, he had assayed the effect of the venom's antiserum and seen that it inhibited the formation of fibrillar halos in vitro. This result led him to test the effects of the specific antiserum to the salivary NGF prepared using standard immunological methods. When it turned out that it too inhibited the formation of the halo around ganglia cultured in the presence of salivary NGF, I proceeded to inject small quantities into newborn rodents. The inspection of the sympathetic ganglial chains of one-month-old animals that had undergone the treatment dumbfounded us. Once again, there is no other term to describe our amazement in the face of yet another utterly unforseeable effect bordering on the miraculous. The superior cervical ganglia—which in mice and rats of that age appear to the naked eye as small ovoidal formations on the sides of the two cervical carotids—had been reduced to the size of minuscule nodules barely perceptible even at high magnifications. The same size-reduction characterized all of the other ganglia along the paravertebral and prevertebral chains as well. Microscopic examinations revealed that sympathetic neurons had suffered almost total atrophy.

The destruction of the sympathetic system in no way affected the normal development of other systems or interfered with the normal somatic development of the rodents. Viktor, who happened to come into the lab just as we were observing the first evidence to this effect, reacted with enthusiasm. "Remember the date of this finding . . ." he said. "It marks a memorable event in neuroembryology." It was 11 June 1959. Several years had to pass before we understood fully the significance of this phenomenon—which dramatically demonstrated the fundamental role NGF plays in differentiation and in the very survival of sympathetic cells—and before it was found that sensory nerve cells, other nerve cells in the central nervous system, and one cell line of the immune system as well depend on NGF for their differentiation and function.

Before leaving Washington University, Stan discovered a phenomenon that in his hands was destined to become a magic wand that opened a whole new horizon to biological studies. The phenomenon, which none of the "smart guys" he admired would probably have paid any attention to, consisted in the precocious opening of the eyes of mice injected with a salivary NGF that had not yet been entirely purified. Stan discovered that the effect was due to the precocious growth of the epidermic strata of the eyelids which caused their two edges to separate one week earlier than those of controls. The contaminating factor was the protein molecule Stan proceeded to purify and called Epidermal Growth Factor, a name kept even after it was discovered that it also has a powerful proliferative effect on connective tissues. Recent developments in this line of research have revealed extremely important connections between the mechanisms governing the proliferation of normal tissues and those causing neoplastic growth.

One hot July evening of that same year, at sunset, Stan and I parted. Our parting took place with the same simplicity that had characterized our first encounter upon my return from Brazil six and a half years previously. He set out with his limping step, smoking his pipe, Smog as always at his side. I followed them from the window while thinking back with gratitude on the years spent together and looking forward with anxiety to those I would have to face without his precious aid.*

---

*By an unhoped-for turn of events, Piero Angeletti, a young Italian doctor from Perugia who was working at Washington University Medical School, had only a few months before Stan's departure joined our group to study the nature and mechanism of action of NGF. Thanks to his excellent preparation in biochemistry, Piero was able to take over for Stan the investigation of the structure of NGF, while at the same time working with me to analyze its spectrum of action. I will touch upon the results of our collaboration in chapter 22.

CHAPTER 20

In Memory of
a Friend

Bred to a harder thing
Than Triumph, turn away
And like a laughing string
Whereon mad fingers play
Amid a place of stone,
Be secret and exult.

—W. B. YEATS,
"To a Friend Whose Work Has Come to Nothing"

THE PERSONS of one's own or of the opposite sex, of similar or different social rank, profession, and age, whom one meets in the course of a life—especially when one has, as I have, lived longer than three quarters of a century—can easily number in the several hundred thousands. Among these, perhaps fewer than a thousand leave permanent traces in one's memory, and few of these are colored by affective tonalities. Extracting only those who have been "real encounters," people who have influenced in one way or another the course of both one's thoughts and one's life, brings the number down to a few hundred or—according to more rigorous criteria—to a few dozen. In relation to these people, there remains, even after many years or even for the duration of an entire life, the indelible recollection of the moment and the occasion of the first encounter, of the ones that followed that made one aware of the significance of the relationship, and finally the moment and manner—if it has already taken place—of the last farewell. None of the qualifications of sex, rank, profession, and age influence in the least these "real encounters."

A New Life

This, at least, is the conclusion I come to as I review my experience. In my life, which I consider to have been particularly fortunate in terms of number of encounters—as a result of my having lived for thirty years partly in Europe and partly in North America and of having met, by virtue of my profession, illustrious figures in the worlds of science and politics and in the *beau monde,* as well as people of less privileged origin and position—the "real encounters" have been entirely random and unpredictable. Trusting my memory, I never developed the habit—nor do I regret not having done so—of keeping any kind of record, still less a diary, because I believe that, if memory has not taken the indelible imprint of a given event, then it could not and should not be brought back to life by mere written witness. I believe, in fact, that the very act of recording an event causes, if only unconsciously, a distortion resulting from the blatant desire of the diarist to make use of it as an account to be exhibited to third parties, as a way of reliving in old age a particular moment, and of making one's descendants partake in it or even, if one is especially vain, for its value to posterity.

Of Fernando, known to his friends and to me by the diminutive Nando, nothing remains to me other than his signature at the bottom of a snapshot taken in April 1953, on the day of his departure for Chile. The photograph, snapped a few minutes before he boarded a Greyhound bus for New Orleans where he was to embark on a cargo ship bound for South America, is on my desk today, and I contemplate it as I contemplate the photographs of other persons dear to me. The image, which came out remarkably clear, shows in three-quarters profile only a youthful face whose intensity of expression has always made me think of the countenance of a beautiful young monk in a painting of Titian's. The picture had been taken because Nando and I both believed that his return to Santiago was to be final, but events took a different turn, and he came back four years later to take final leave of his friends and of life.

Our first meeting, on a wintry morning in 1951, had been preceded by a phone call from George Bishop, the neurophysiologist at the medical school of whom I have spoken. The call was for Viktor, but since he was busy teaching, I answered it. As usual I had great difficulty understanding what was being said on the other end of the line, a difficulty increased not so much by Bishop's accent and brusque way of speaking as by the pipe clamped between his teeth

170

which distorted his every syllable. "There is a chap here from Santiago de Chile," I managed to decipher. "He's been sent to me by Luke, the well-known neurophysiologist at the Catholic University there. The chap's name is Fernando, and he seems quite bright. He would like to meet Hamburger. Will he be around at noontime?"

Fernando arrived punctually at noon, and Viktor, having finished his lesson, received him in his study and called me in to meet him. From the start, I was struck by the extremely intense expression of his very black and large eyes. His hair was dark chestnut in color, thrown back from a high forehead, above thick eyebrows knotted tightly in the effort to express himself correctly in his Chilean–Spanish English. His shortness and robust muscular frame suggested a young athlete or a bullfighter. He was surprised at Viktor's invitation to join us for lunch at the nearby campus cafeteria, and refused: "I'm sorry, but I never eat lunch. I prefer to pass the time in the library." Viktor smiled at his youthful ardor. Nando was twenty-nine and had time to study and nourish himself like other common mortals, but he stood firm. We accompanied him to the library which for him, as it had been for me, was something to wonder at, both of us coming from countries where austere shelves of nineteenth-century publications dominated and the number of recent periodicals was minimal. He buried himself in the periodicals and resumed his conversation with us after we had returned from the cafeteria. He knew Hamburger by reputation and had hoped, in coming to St. Louis, to take advantage not only of Bishop's renowned competence in electrophysiology but also of Viktor's in experimental neuroembryology.

The intention of frequenting both institutes simultaneously turned out not to be practicable, but the mutual affection and affinity of interests—beyond the merely scientific—that existed between him, Viktor, and myself, led him to frequent our lab in order to exchange ideas on various subjects. While the research he was engaged in with George Bishop, James O'Leary, and Bill Landau at the medical school was proceeding well, Nando, in the fullness of his youth and the universality of his interests, felt the need to exchange ideas with Viktor and me who were, owing to our European origin, more in tune, he felt, with his way of thinking than were the all-American members of Bishop's team. Nando was therefore also very happy to come to dinner at my house sometimes, where we would bring each other up to date on our most recent research.

In reality, it was I who brought him up to date on my research more than he did me on his. I was then in the initial and exciting phase of my research on how grafts from a malignant mouse tumor affect the sensory and sympathetic nervous cells of chick embryos. Such was my enthusiasm that day and night my every thought was concentrated on the phenomenon. I felt as if I, too, was a carrier of the tumor and subject to its prodigious effect, but flattered myself by thinking that the luxuriant proliferation of fibers sprang, in my case, not from sensory or sympathetic cells but from the cerebral hemispheres to which are attributed far more precious faculties. If not nervous fibers, thoughts certainly were budding in my brain, and so tumultuously as to leave no time for pursuing other ideas. Thus, no sooner had I asked Nando about his results than, giving him no time to answer, I would remember that I had not yet told him of a new fact discovered that morning: a fact not only confirming the hypothesis formulated during the previous days but also pointing to new and exciting paths to follow. In this respect—if perhaps only in it—my behavior was similar to that which Dylan Thomas attributed to himself as a constant in his relationships with friends. As soon as he met a potential new friend, he would ask him to tell him some episode of his youth, but if the friend were not quick, Thomas would overwhelm him with innumerable stories from his own childhood.

Nando would listen, and it seemed to me he did so not only out of kindness but because he was fascinated by this story of mine. He took his revenge, however, whenever the conversation turned to music or adventures, in which his childhood and adolescence had been much richer than mine. In matters of music, he was an enthusiast and a great authority, out of natural talent and because he had been exposed to it since early childhood. His brother—his senior, I believe, by a few years—was a well-known and highly esteemed composer. Nando's stepfather, whom he loved and venerated, was a famous composer. His mother also had a strong musical talent. If to Guido I owe my initiation into music in my university years, and the intense enjoyment under his guidance of Mozart, Beethoven, Schubert, Chopin and the other great composers of the past two centuries, to Nando I owe my acquaintance, which soon turned into passion, with Bach and his predecessors from Praetorius to Buxtehude, to the Italian musicians of the seventeenth century, to the composers of ancient liturgical music such as had in the early 1950s only recently appeared on records.

In Memory of a Friend

Taking advantage of what I deemed Nando's unlimited musical culture, I spent long hours of the weekend with him in the store of our music supplier, Frances Rooz. In those days, almost my entire month's salary passed from my hands into hers. Following Nando's advice I bought the best hi-fi set I could afford; and from then on, as soon as I arrived home in the evening, before preparing myself a frugal dinner, I listened to one of my favorite records. Using one of those little gadgets that starts the record player automatically, making it a sort of alarm clock, I'd pile on a series of records—something people no longer do today because the advantage of having records loaded automatically is set against the cost of bad record maintenance—and then turn the volume all the way up. I didn't worry, as perhaps I should have, about my neighbors. My dear friend Arthur Kornberg thought this habit of mine very inconsiderate toward the unfortunates living next door or above or below me. "At least," he suggested, "pin on your door the next morning's program. That way they won't have to trouble themselves trying to figure out what composer you're playing." But Americans are in this regard more tolerant and kind than Europeans. No one complained and I continued the habit until, many years later, I left my apartment and Washington University for good to return to Italy.

Another of Nando's passions was for Nature, which he enjoyed not out of a deep knowledge of geology and botany, as was the case with Viktor, but for its beauties. From what he told me, Nando had had the good fortune to be born in a country not inferior to Italy in terms of natural beauty, though the range—from the north, where vegetation is subtropical, to the Tierra del Fuego close to the South Pole—is very different. He knew and deeply loved this country of his which enjoys the unique privilege of lying, for its entire length, between mountains and oceans. In less than an hour, he could go from Santiago to the snow-covered mountaintops that rise near the city. He was equally attracted by the deserts in the north which had at one time represented the country's wealth owing to the great nitrite deposits created by billions of birds leaving their excrement there through the centuries. This natural wealth had recently lost its value because of the production of chemical fertilizers in the United States. In the course of frequent trips up north, he explored the desert regions with natural caves, which he entered in order to see, hanging from the walls, the millions of vampire bats that inhabit them. Since not many years had passed since the Nazi-Fascists' ferocious anti-Semitic cam-

173

paign had rendered the term *Jew* synonymous with *vampire*—Jews being, according to the press of that time, as thirsty as vampires for the blood of their victims—I found it difficult to imagine liking the little flying mammals. Furthermore, their appearance, which is even more repulsive than that of their relatives, the normal bats inhabiting Italy's caves, had caused them since the Middle Ages to be the object of sinister legends and, in recent times, the protagonists of horror films. Vampire bats, as we know, are rare among mammals in being able to fly. This ability they owe to the transformation of their hind limbs—humerus, radius, and four toes—into wings or, rather, into membranes connecting the minuscule bones much as fabric connects the ribs of an umbrella. Only the big toe, or thumb, is free and has been transformed into a claw which other kinds of bats and specimens of *Cheiroptera* use as a hook to attach themselves to the walls of caves and, in the case of vampire bats, to a victim's skin. Vampire bats owe some of the dislike they inspire to their great ugliness, the result of cutaneous excrescences on the extremity of the snout, and, even more, to the very long, razor-sharp incisors with which they puncture the skin of their prey to suck their blood. Nando, however, who knew these bats well from his explorations of the caves, had developed a strong liking for, and a sort of solidarity with, them—not least as a form of protest against their terrible reputation. One night, tired out by his wanderings, he had fallen asleep in a cave. Upon waking, he noticed that a vampire bat, attracted by the warmth of his body, had nestled in the cavity of his armpit and, from inside its little refuge, was peering at him with eyes that were intelligent and full of the sense of security and trust that a child has contemplating his sleeping mother. Since then, Nando's friendship for vampire bats had prevented him from using them as laboratory specimens.

In the spring of 1951, Nando's wife arrived. She seemed to me even more striking than he had described her. Her hair, shining like the wings of a swallow, the brilliance of her eyes, and the amber color of her very handsome face gave her a beauty typically Latin. She had come with their three children. The eldest, a boy, had a proud demeanor and a curiously protective attitude—not unlike that of his father—in regard to his mother. Both, in fact, showed signs of a macho complex which conflicted with the anything but submissive and rather mocking manner of their wife and mother. The two girls, eager and open-hearted had come out onto the balcony of life too recently to appreciate its value and fear its snares. A year later, his wife decided to go back to Santiago with the three children ahead of Nando, who joined her there in 1953 after finishing his research with Bishop and his collaborators.

Nando was sorry to leave St. Louis, but it would not have been possible for him to live apart from his family, nor was there at that time any possibility of his finding a permanent position at the medical school. I accompanied him to the Greyhound Bus station and bought him a knapsack to carry all the toys he had purchased for the children.

A few days before, he had begged me to take over his gigantic Ford, a pre-war model, assuring me that, in spite of all appearances, it was in excellent condition. The only small problem it had—according to him—was that the hand brake had long been broken. He assured me that it wasn't worth it for me to spend time and money trying to have it fixed. As for driving the car—I had never touched a steering wheel in my life—it was child's play; even his son had learned how to handle it right away. I followed his advice. In those days in Missouri, all that one needed to get a driving license was to send or go in person to pay the extremely reasonable sum of twenty-five cents. One didn't have to show a document proving that one was able to drive, or take any exam.

I had driven the first time with Nando sitting next to me, and managed without the slightest difficulty. I was to have some, however, a few days after his departure. I had noticed that I had to press the foot brake very forcefully in order to get the antediluvian Ford to stop. I decided to see a mechanic the next day to have him change the oil and check the brakes. At that time I was living where I was to spend all the years of my stay in St. Louis, in a faculty and student apartment building situated on top of a hill. From there one descended a rather steep slope, skirting one of the walls of the building, and, at the end of the slope, took a sharp turn into a wide boulevard, the principal artery of University City where Washington University is located. During my descent in the Ford the heavy car immediately gathered speed. Entirely in vain were all my attempts to stop it by stamping madly upon the brake pedal. The last drop of oil having evaporated, the brakes had ceased functioning altogether. Knowing that I couldn't make use of the hand brake, and having lost all hope of stopping the car's career, I thought that my last hour had come—as well as that of anyone unfortunate enough to be in my path. With all the sangfroid I was able to muster, I managed to take the boulevard at full speed, and miraculously avoided collision with oncoming cars. Paying no heed to the cries of rage and abuse that assailed me, I concentrated all my attention on maintaining control of the steering wheel, and succeeded in moving the car to the side of the road. Fortunately the boulevard was level and flanked by unfenced lawns. Left to its own inertia and with no fuel to feed it, the Ford slowly lost momentum and came to a halt on the grass, but not without traveling a good hundred yards. I got out paler than the dead, clenching my teeth, and muttering, "Nando, how could you do this to me?" But Nando was at the time already at sea on his cargo boat, probably dreaming about his vampire bats. With the help of some passersby, I called a towtruck. From that moment, I decided that the Ford might find its place in a museum but would never be driven again. As a lesson in driving, however, the experience was excellent. From then until today, over thirty-five years later, I have never had another accident, not even in the chaotic traffic of Rome.

Neither Nando nor I were in the habit of writing letters, and exchanges between us became ever more infrequent. One day in 1956, however, I received a letter from him expressing a great desire to return to St. Louis, there to resume the research he had interrupted and had been unable to continue with any success in Santiago. The situation was now more favorable for him to return. Having unexpectedly been awarded a sizable grant, I could fulfill my dream of integrating research in neuroembryology with research of an electrophysiological nature in our own laboratory, without having to have recourse to the collaboration of others. Nando was the perfect person to carry out these experiments with us. As Viktor was entirely of the same view, we were able to offer Nando a modest stipend out of my grant funds and the position of research associate. He accepted enthusiastically and, in May 1957, reappeared at Rebstock Hall, where Viktor and I greeted him. Though he was the Nando of old, full of enthusiasm and dreams, something about him seemed different. His eyes were shiny as if he were feverish, and his hands trembled slightly. I attributed these symptoms to his happiness at being back together with friends and at the prospect of resuming the conversations and the listening to music we had both so greatly enjoyed. After a few days, however, I became alarmed by his frequent desire to

176

"clear his thoughts" with little glasses of whisky or other drink. When I asked him whether this had become a habit with him, he assured me that I shouldn't worry, that it was a way for him to think more clearly and to overcome bothersome worries. Though he adored his wife, it seemed that the harmony of old no longer existed between them. He appeared—as I had the opportunity to notice when she and the three children joined him a few weeks later—irascible and impatient. A fourth pregnancy was, however, under way, and the birth of a baby girl two months later seemed to bring joy and peace back to the family.

Nando turned enthusiastically to designing and building a small copper-insulated unit for electrophysiological experiments, complete with all the instrumentation necessary to record the intracellular and extracellular potentials of nerve cells under investigation. As object of study, he preferred fully grown animals to chick embryos, with which he felt insufficiently familiar to work well. Ruling out mammals because of his unwillingness to inflict suffering upon animals that he loved as much as he loved cats and dogs, he opted for turtles instead. They had the advantage of possessing a less complex nervous system, even if certain circuits, such as the olfactory one which he was planning to analyze, did not differ substantially from those of mammals. Their expressive atony, common to all reptiles, and their awkward way of moving, like that of their gigantic predecessors, help human hypocrisy not to attribute emotive capacities to them and to doubt whether they actually experience sensations of pain.

Having thus dubiously set his conscience at ease, Nando could dedicate himself without anxiety to the opening of a turtle's skull and to the analysis of the neuronal circuits of the olfactory system. He showed excellent technical skill in these experiments and, what pleased me even more, a lively interest and the capacity to work intensely and uninterruptedly for long hours. These positive facts gave me the illusion that he had found once again the peace of mind I had noticed him to be lacking. He was cheerful and full of vitality when in the evenings we listened to our favorite records or when he spoke to me of friends to whom he was deeply attached and had had to leave in order to come back to St. Louis. After his return, the records he often asked me to put on were Bach cantatas, especially those full of a sense of death. His favorite record was "Die Winterreise," Schubert's song of the lonely wanderer in winter. This music was complemented by that marvelous record in which Dylan Thomas recorded, in his beautiful, cavernous voice, the story of his life and of his downfall: "When I was a man you could call a man. ... When I was a half of the man I was. ... Now I am a man no more no more/And a black reward for a roaring life." This had become by now Nando's dominant thought: his ineluctable degradation and unrestrainable fall.

Nando was also beginning to develop worrisome physical similarities to Dylan Thomas. His face had lost its youthful freshness, and his gaze, though at times as intense as ever, would suddenly become cloudy, and a web of dilated capillaries would stand out against the white of his eyes. His hands, as I had immediately noticed upon his arrival, trembled; and early in the morning, his breath smelled of whisky. He had accepted my request that he hold a series of lessons on neurophysiology in my course, and I was assailed by anxiety when I noticed that he easily lost track of what he was saying and that his trembling was evident as he traced graphs on the blackboard. His relationship with his wife became gradually more tense in the coming year. Nando complained that she was very impatient with him and openly showed lack of both tolerance and confidence in his regard. I witnessed with deep distress the thickening clouds that hovered over the future of the young couple, though their four children were blooming and becoming every day more beautiful. The boy, who was then nine years old, seemed to assume an ever more protective attitude in regard to his mother.

Everybody at the institute was by then aware of Nando's chronic alcoholism, nor did he try to hide it any longer. He was always going off to the men's room, from which he would emerge more cheerful, never seemingly drunk, as if refreshed and ready to resume work. A little bottle of whisky always bulged inside his coat pocket, though he did attempt to conceal it.

In the summer of 1959, his wife returned to Chile with the three girls. Although Nando and the boy lived in my apartment building, I confess that I abstained from going to visit them. In spite of my affection for both, I did not feel like playing a maternal role toward them, never having had any inclination for such a role.

One day in the middle of winter, the boy called me and said that his father, whom I by that time saw only rarely, had not come home that evening. Nor did he return the following two evenings. Full of worry and fearing the worst, Viktor and I thought of asking the police to help. Nando reappeared, however, on the fourth day, in a hardly recognizable state. He had wandered about for four days in the cold of February, believing that complete solitude, meditation and abstinence from drinking—which he claimed to have achieved—were his only remedy. I begged him to turn to a specialist, but he shook his head. Either he cured himself, or no one else could do it. And no one else did. This marked the beginning of the end for Nando.

He came to the laboratory, which he had not visited in a while, and his eyes were feverish. I begged him to go back home immediately, given the obvious pulmonary difficulties he was having, his breathing short and irregular, and promised him I would visit him in the evening along with Piero Angeletti. Nando greeted us, happy at our visit. Having taken some antipyretic drugs and tranquilizers, he assured me that he felt much better. So much better, he said, that the next morning he would show up at seven to do an experiment he had been planning in the last few days. I told him that he was clearly not at all better, with his burning hands and feverish eyes, and begged him to check into a clinic the following morning for the cure he so desperately needed. He protested that he felt very well. We left at ten o'clock. I was more worried for him than ever. As soon as I got home, the phone rang. It was Nando. He wanted to express his gratitude for the visit and reaffirm his conviction that this time he was on the road to recovery, that he would never have another relapse. Deeply moved, I thanked him and wished him a good night.

At six in the morning, the phone rang again. I thought it would be Nando. I heard instead his son's voice stirred with emotion. "Rita," he mumbled between sobs, "come quick! Quick." He didn't know what had happened, but "Daddy looks so strange." I ran upstairs to their apartment. Nando was lying on the ground next to the phone, his eyes wide open and his arms stretched out. Hoping to revive him, the child had poured water on him, which had wet the carpet around his body. The body itself was rigid and cold, and it was clear that death had occurred several hours before. Given his position with respect to the telephone, I believe that death struck him just as he was putting down the receiver after speaking to me. I called the janitor and Piero Angeletti, and we informed the police right away. They arrived a few minutes later, two heavy-set policemen along with an ambulance. They looked at Nando with expert eye. "Was he a drunkard?" they asked me. I nodded slightly in affirmation, hurt by the brutality of such a definition, which somehow did not suit Nando, while his son kept close to my side and remained silent.

The larger of the two policemen put a hairy hand on his shoulder. "Son," he said, "this is the time to prove you're a man, and not a boy any more." The boy did not reply but said that he had to get ready so as not to be late for school. We told him that he could stay home that day if he wished, and then realized that it was Washington's Birthday, a school holiday. We called some friends with children of his age who came and took him to their house.

The funeral took place the next day at the Unitarian Church in Washington University. Present were George Bishop, gloomy and frowning, Viktor Hamburger, and most of the members of the medical school's department of neurophysiology and of our department as well. I asked that Nando's favorite Bach cantatas be played. Everybody had loved him, but I think that I more than anyone could appreciate his exceptional sensibility and the unusual vividness of his moral and intellectual qualities as I knew them. For this reason I could not forgive myself for not having succeeded in saving him. As for his son, such was his pride that he succeeded in being "a man" as the policeman had said. I never saw him again. For a few years, he wrote to me and sent me photographs of himself, his mother and sisters, but with time his letters became more infrequent. In describing Nando, I have tried to be faithful to my recollection of him. And if his children ever read these pages, I hope that for them his splendid figure may live again. "Bred to a harder thing than triumph," he turned his back on success in respect to his studies of the neuronal circuits of the turtle's olfactory system. In spite of his lively intelligence and creative capacity, he left no significant contributions to neurophysiology. During his life, which ended on that February night in 1960, just before his thirty-eighth birthday, he loved everything with a passion: Buxtehude, Bach, Schubert, cats and vampire bats, nights in the desert and on the tops of mountains under the starlit sky.

His ashes are kept in a small urn on which is inscribed his name. It lies on the top row along with many other urns in a wall of a Gothic-style chapel, in midtown St. Louis. The rays of the sun, which in the evening filter through the great windows in the opposite wall, shine on his name. I have often visited the chapel. In the spring, one can look out the windows and see branches covered with buds and hear the songs of birds. I never discussed the question of death or an afterlife with Nando, but his presence was on the wane when he was among us, especially in the last period, and I have a feeling that his "afterlife" had then already begun. I think that he felt the nearness of his end—provoked apparently not only by alcohol but by the concurrent use of large doses of strong sedatives—yet did not fear it; on the contrary—in this, as in many other respects, a follower of Freud's—he thought death in reality an unconscious and overwhelming aspiration of man. For Nando, in any case, it was only one of the infinite aspects of life, and he did not fear death in the same way that he did not fear the vampire bats he so loved.

PART IV

Back to My Native Country

CHAPTER 21

The Tug of
Early Affections

LATE ONE EVENING in January 1961, in Detroit, I met the "foundry owner," the character impersonated by the actor Ermete Zacconi who had elicited in me as a child strong feelings of rebellion against the opposite sex. This man in Detroit had in common with that character not only profession but also corpulence and success. He had financed the scientific symposium I was attending. At the end of the series of meetings, I and some other participants were invited to his sumptuous mansion. After dinner we followed him down to the basement where he had built a miniature reproduction of one of the great railway centers of the East Coast. With the glee of a child, he operated dozens of switches and exchanges, maneuvering trains amid flashing lights and sirens through an intricate network of rail lines that extended from one room to another.

In my early passion for trains, I had as a child spent many hours on the footbridge near Turin's Central Station where I observed the intense traffic of locomotives and railway cars while dreaming of the unknown countries whence they came and to which they were going. I dreamed of the "great trip" Father had promised to take the family on when the "little girls," as he used to call Paola and me, reached the age of thirteen. It was to be a fabulous journey on the Trans-Siberian railroad, across a Russia as yet unscathed by revolution. From that train, the walls of whose passenger cars were all see-through crystal—according to Father, who had never seen it—we would be able to watch polar bears amble over boundless flat lands covered with snow.

BACK TO MY NATIVE COUNTRY

The trains of the foundry owner from Detroit awakened in me the recollection of those childhood fantasies and made me nostalgic for a past that, though distant, was still so alive in memory. Would I spend the rest of my life in that marvelous isolation offered me by the Midwest—an isolation interrupted only by the one month each summer that I spent with my relatives in Italy? Now, after many years away, I felt a strong desire to be closer to Mother and Paola, and to Gino, Nina, and their families. During the fifteen years we had lived so far apart, we had at least partially made up for the separation by means of a voluminous correspondence: in which I told Paola everything about my life, and she hid from me everything about hers, filling the pages of airmail stationery with news on the health of our dear ones. But, behind her silence, I could feel pulsating the intensity of her life; nor was I fooled by the serenity that seemed to transpire from her words. Her creativity, like that of all genuine artists, is the filtered product of an inner experience that consumes those who have the rare privilege of tasting it. Unlike scientists, who are gratified by discoveries that reveal—in the most fortunate cases—only infinitesimal fractions of the world surrounding them, artists aspire to formulate and transmit their own conception of the world, an aspiration that allows no pause or relief to be gained from one's accomplishments, which fall always short of the goal one has set oneself.

In his Introduction to a 1939 monograph about her, De Chirico wrote:

> Paola Levi-Montalcini is gifted with great pictoric temperament. . . . It is the ascending movement in the landscape "The Olive Trees" that engenders the painting's poetic motif: the displaced trunks, the bushes, everything rises like heavy fumes, like the elongated figures and the tormented towns of El Greco. . . . In the landscape entitled "Hill," firm and right touches powerfully model the ground that rises up to the line of the sky. This landscape, one of the most suggestive by the young artist, recalls Eugène Delacroix's "Pyrenees Landscape." . . . When Paola Levi-Montalcini depicts the human figure, it is always its aspect as an apparition that tempts her. Her still lifes make one think of a separate life, a mysterious life of fruit and objects . . . that life which in the midst of the world, in the midst of the *other life,* exists in its own closed circle and does not appear other than to real artists and real poets.

Each time I returned to Italy, I contemplated the compositions she had come up with during Turin's long foggy winters. I was amazed

186

The Tug of Early Affections

by the incisive power and intensity of her "marks"; but in response to my enthusiastic exclamations, she would only shake her head. Now in January 1961, I wondered whether returning to live in Italy would afford me a way of enjoying being near to and communicating with her. Or would she have shielded herself as she had since childhood, so that, though thriving from the deep affective ties binding us to one another, I would not be allowed access to those realms from which she derived her creative energies?

To my great relief, I did succeed, at least in part, in overcoming the barrier when, after the death of our mother in 1963, Paola came to live with me in Rome. I was thus able to watch come to life the kinetic-luminous structures, watch them rise from the floor like stalactites; and to see the large copper and steel engravings in which letters of archaic alphabets, caught in the folds of Archimedean spirals, send indecipherable messages, mathematical symbols that meander—shattered the bonds in which they are usually constrained—freely in the void; and the profiles of marine birds liberated in interplanetary space. Our conversations, which began in her studio, have continued in the art galleries and museums we visit together in European capitals and in the United States during annual summer or fall pilgrimages. These occasions always serve to reveal Paola's exceptional sensibility, inexhaustible enthusiasm, and sure intuition in the discovery of young, new talent—which most people fail to notice, being lost amid the host of those who are better salesmen of themselves, and more successful; and reveal also the similarity between our two life itineraries, despite the differences between our choices and the diametrically opposed rhythms of our daily activity. Since adolescence, both of us ruled out the idea of creating families for ourselves, deeming such responsibilities to be hardly compatible with full-time dedication to the activities we had elected. Neither she nor I have ever regretted our choice. A surprising analogy between our two lives became manifest also in the relationships we had with our teachers. Her relationship with Felice Casorati was practically identical to my own with Giuseppe Levi. Hers, as mine, was dictated by a profound admiration and affection which did not, however, hinder our search for an identity of our own.

My return to live in Italy, which I had been dreaming about for many years, became reality, two years before Mother's death, in a way gratifyingly in tune with my wishes. After meeting the foundry owner, I told Viktor of my desire to spend a longer period in Italy than my usual one-month vacation. The previous year, in 1960, I had given a conference on the Nerve Growth Factor which had aroused considerable interest in certain Italian scientific circles. It was on that occasion that I had first aired the idea of establishing a research unit in Italy. But would I be able to reconcile such a plan with my position as full professor at Washington University? Viktor showed himself to be, as always, understanding and ready to help me. I would be allowed to give my yearly semester-long course in neurobiology to my undergraduate students; and, during the quarter that I planned to spend in Italy, my young and able collaborator Pietro Angeletti would take over the direction of the laboratory.

To the dean of the university, Viktor explained my desire to spend three months out of the year in Italy in order to establish there a small research unit which would collaborate with Washington University. The dean approved my request; and, with his endorsement in hand, I went to Washington, D.C., to submit my proposal to the National Science Foundation. They quickly presented me with a grant, and I returned to Italy in the spring of 1961 to set up in Rome the research unit which I shall discuss in the next chapter. I now began to stay half the year in Italy, allowing me the great joy of spending every weekend that I was there with Mother in the last two years of her life.

The affective ties between Mother and me were of equal intensity yet very different from those that existed and still exist between Paola and myself, and that separation and the passing of the years have strengthened rather than weakened. Neither Mother nor I, however, were ever tempted to reveal the nature of our relationship in an effusive manner; and our reciprocal affection manifested itself more through silence than through words. Our bond had been the cause of unmentionable anguish for me during the years of persecution, when I was terrified at the thought that she might fall into Nazi hands. She, instead, was not troubled in the slightest by the threat looming over us. Her courage and serenity in those critical moments were the expression of her nature, and also of an assurance that, whatever might happen, we would suffer it together. Both by her and our good fortune, we managed to escape the SS—a feat all the more miraculous given the naïveté, which bordered on blindness, of our plans to avoid capture. When I received Viktor's invitation to go to St. Louis, it had been Mother who encouraged me to take the step, knowing how much

The Tug of Early Affections

I desired to live such an experience in the United States; and the following year, it was she again who urged me to accept the offer to stay there. In the hundreds of letters she sent regularly each week during the many years we were apart—letters that I zealously conserve—there was never the slightest mention of the serious asthma attacks that had begun to afflict her.

In June 1963, while in St. Louis, I received news that she had suffered a fall and fractured a thighbone. As was common practice in those years, a peg had been inserted to facilitate the joining of the bone segments. Upon returning to Turin in July, I found her convalescent and forced to use a wheelchair. She improved and was able even to visit me in Rome, for the first and last time, in September. Toward the end of the month, we returned to Turin for the removal of the peg in her thighbone—an operation more difficult and painful than foreseen. In the following days, Gino, Nina, Paola, and I took turns at her bedside, troubled by her asthma attacks and by a surge in temperature which refused to wane. On 3 October, after a pause at home, I went back with Paola to the hospital. It was three o'clock in the afternoon when, suddenly, we saw three elderly nuns appear from behind a colonnade in the hospital's large entrance hall. All were dressed in black, their faces framed by veils with vast starched brims rising on the sides like the wings of butterflies. In the penumbral darkness of the hall, their sudden appearance and slow, solemn gait caused a shiver to run down my spine. Squeezing Paola's hand, I murmured, "The three Fates!" Paola said she had had the same thought. Quickening our pace, we reached Mother's room, and found her still in pain, but smiling. Seated at her side, with her beautiful slender hand in mine, I felt reassured.

At the end of the week, her condition had improved. Thus, my mind set at ease by the doctor, I left for Rome where I planned to stay for two days: a new researcher had just joined our group, and I felt it my duty to see that she was "broken in." Was it really a good enough reason for me to leave Mother, even if only for a couple of days? I would pose myself the question an infinite number of times during subsequent days, months and years, tormented by my impotence to change the course of events. The following evening at seven, I answered the phone to hear Paola's broken voice: "Mother——." She did not have to add other words. I understood. I returned immediately to Turin and was met at the station by Gino, Nina, and Paola. Upon arriving at the hospital the previous afternoon, Paola had been greeted by the open arms of a nun who said, "Be courageous and accept the will of the Lord." The three Fates had preceded Paola. Had Mother seen them or intuited their presence? She had turned pale, the glass of water she had requested of the nurse had slipped out of her hand, and she had rested her head on the pillow.

When, a few hours later, I held her deathly pale hand in my own, I contemplated through my tears her splendid countenance which, in its diaphanous immobility, made me think of the saints on fourteenth-century sarcophagi, and kept repeating to myself that nothing could ever, until the end of my own life, alter the ties that had bound us since my birth. She herself, when I left Italy each year after my brief summer stays, encouraged me to face our separations serenely, in the same way as I would some day courageously have to confront our definitive parting. "When you receive the news," she would tell me, "take the first plane you can. Not so as to get here on time to say any last farewell or to hug me one last time, but for Paola's sake, who in that moment will need you beside her." That day had come; and though I tried to find comfort in the thought that death had come stealthily and been sweet for her, I felt as if the thread of my own life had also been cut. Every night I saw her beloved face again and held her hands tight in my own, saying, "So, thank God, it was only a nightmare. We are together again." But it is a painful and well-known fact that the solace afforded us by dreams, when they bring back to life people dear to us, is never complete, and it is even less lasting. In our desire not to wake up, which makes us struggle to delay that deeply feared moment, we sense how tenuous and illusory the wall is behind which we have sought refuge. And every morning we experience the same suffering, which we know no artifice will ever help us elude. Nor, in reality, do we desire as much, since it would separate us even further from the person we loved.

Two days later, on a warm October day, we accompanied Mother to her last resting place, where Father had been lying for thirty-two years. The tomb, which Gino designed, is a granite wall with a rough, ivory-colored surface, and inscribed with the verses from Ecclesiastes that, thousands of years after they were first proffered, express the austere and desperate wisdom that inspired our forefathers: "What profit hath a man of all his labour which he taketh under the sun?" On the marble which covers her tomb a brief epitaph is

190

sculpted, summarizing the itinerary of her journey: "She left small things for big ones, then, suddenly, she left all things for peace." The family cares absorbing her in the years of her youth and early maturity had been followed by other problems, which she had faced serenely and with increasing detachment from daily worries. Finally, peace had come, the seal of every life, be it stormy as my father's or as sweet as hers.

CHAPTER 22

The Miracle of
Maxwell's Demon

IN 1983, Robert Walgate published—in *Nature,* the renowned scientific review for which he was a correspondent—an analysis that showed both benevolence and a deep understanding of the facts in the case regarding the current state of research in Italy. He resorted to an efficacious comparison: thus, where the demon of James Clark Maxwell, the Scottish physicist, was unable to separate high-velocity molecules from low-velocity ones in a gaseous mixture by allowing the former to pass through a small hole in the dividing wall in the container—a separation that would have violated the Second Law of Thermodynamics—Italian scientists have shown themselves to be sometimes capable of doing some "high-velocity" scientific work in spite of the structural financial and political chaos in which they operate. How to explain such a miracle? I shall attempt to provide an answer based on my experience during the last twenty-four years— that is, since the time when I realized my dream of setting up a research unit in Italy. Having experienced the situation from the inside, both as participant and as spectator, I carry with me the baggage of experiences only a "native" can possess. Furthermore, knowledge of how the same problems are handled in the United States puts me in a position to evaluate the positive and negative aspects of two very different strategies and systems.

At the beginning of the 1960s in Europe, and in Italy perhaps even more than in other countries, economic recovery had been much faster than anyone could have hoped. But in the scientific realm, severely impoverished after twenty years during which culture had been totally disregarded, recovery proceeded more slowly. In the field of biology, and of genetics in particular, the hearth had been kept alive even through the darkest years by a group of young scientists, headed by Adriano Buzzati-Traverso, who immediately after the war set up the Center of Biophysics at the Institute of Pallanza in Lombardy. The institute, where the mutagenic effects of chemical substances and ionizing radiations were studied, sent its members to international meetings and even organized some in Pavia, where Adriano was named head of the new department of genetics in 1948. After some years spent as a professor at the Seripps Institute at La Jolla, California, Buzzati decided to create an international laboratory in Italy, known as the Laboratorio Internazionale di Genetica e Biofisica (LIGB). Inaugurated in 1962, the LIGB elicited great enthusiasm and consensus among the young but received a cold welcome from the academic community. In Adriano's mind, the new laboratory was to be a place for specialized instruction in molecular biology, genetics, and biophysics—a sort of graduate school; and from the start, it gathered together the most brilliant young biologists and offered hospitality for a few months each year to the founding fathers of molecular biology, including Jacques Monod, Sol Spiegelman, Francis Crick, and Salvador Luria.

The center was soon famous, but not for long. In the late 1960s, the conservative oligarchy managed, with the help of the young revolutionaries of 1968, to cause a crisis that forced Buzzati to resign and the scientific staff to disperse. As a result, not only the LIGB but Italian biology as a whole were seriously crippled.

My project, far more modest than Buzzati's and begun also in 1961–62 in Rome, aroused no opposition in academic circles because of its limited scope and the small number of researchers involved. It was instead welcomed both by the Istituto Superiore di Sanitá (Institute of Health), which provided me with workplace and equipment, and by the National Research Council (Consiglio Nazionale delle Ricerche, or CNR) to which I had turned, after the first year, for limited funding with which to supplement my grant from the National Science Foundation in the United States. Professor Marini Bettolo, director of the biochemistry department of the Istituto Superiore di Sanitá, immediately gave us two rooms in his laboratory and, seeing that they were insufficient, decided to tear down a large marble staircase leading from the first floor, where the laboratory was located, to the basement giving onto the institute gardens. We had available, as a result, three large rooms equipped with all the instruments necessary for structural, ultrastructural, and biochemical research. Within three months I had at my disposal a research unit larger than the one in the department of biology at Washington University.

Thus I began my life as a commuter between two continents, with all the attendant advantages and disadvantages well known to those who, like myself, made a similar life choice around that same time. The highest price I had to pay was that of being separated from friends in the Midwest, from Viktor especially. Over the years, our friendship had been continually strengthened by an uninterrupted exchange of ideas and by the life we shared in the laboratory, where work proceeded in an ideally serene and well-ordered atmosphere. There was also the problem of having to adapt to a certain lack of continuity, divided as I was between two laboratories situated thousands of miles apart, from one another. I was helped to overcome this obstacle by my young colleague and friend, Pietro Angeletti who, like myself, wanted to re-establish links with his family and with research in Italy. We decided to take turns, alternating our presences between Rome and St. Louis, at the head of the two groups which were investigating the same problems, namely the structure and spectrum of action of the NGF.

The most difficult task facing me in Rome was that of personally overseeing the administration of funds and keeping our correspondence in order. In St. Louis, these two responsibilities were part of the duties of the department administrator and a secretary. My lack of organization in such matters had been immediately noticed by Martha Fuhrmann who, in 1960, agreed to take the matter out of my hands, subsequently becoming one of my dearest friends. Martha was a formidable organizer. The first time she entered my office and noticed the mess on my desk, where letters, books, and microscopic samples were randomly stacked one on top of the other, she said in a chiding tone, "Doctor Levi-Montalcini, you did so remarkably well in your work in spite of the fact that you are so incredibly disorganized. Why in the world should you need me to organize you?" When Martha left the job in order to complete her studies, she was replaced by Leonore Friedmann, whose infinite patience and competence carried me through the fifteen years of my divided activity between two continents.

Upon returning to Italy, I also had to adapt to the obsequiousness

The Miracle of Maxwell's Demon

with which employees and young postdoctoral students treated me and other elders. Accustomed as I had become to the cordial "Hi, Doc" of American technicians and students, I was embarrassed by the ceremony which, at the beginning of the 1960s, still regulated relationships between professors and the rank and file. Such rituals, which were abolished with the revolts of the late 1960s, were still very much the norm in the solemn quarters of the Istituto Superiore di Sanitá.

At the level of research, the recruiting of competent persons did not present any problems, so great was the number of young postdocs who wanted to join us. More complicated was the problem of being able to pay them a salary not entirely inadequate to their needs, however modest these might be. Furthermore, there was no guarantee of any continuity of employment, since the payroll depended on the meager funds made available to me year by year. I was immediately relieved to note that the young researchers, who were less worried about the future than their American counterparts and eager, most of all, to test their skills at the lab bench, accepted without hesitation even modest and precarious remuneration. The possibility of taking turns at moving with me to Washington University, so as to become familiar with the latest techniques and tighten the bond between the two work teams, was undoubtedly an incentive. Crossing the Atlantic and spending a few months in the United States was still, at the onset of the 1960s, a privilege enjoyed by only a few. The majority of technicians and young researchers who came to work with me at that time formed, at least up until the 1970s, the nucleus of the group operating in Rome.

My strategy—or, rather, my complete lack of it—during the ten years between 1960 and 1970 was well suited to the casual politics, dictated more by enthusiasm than by any rational program, of the leaders of the CNR at the time. Among the latter, the head of the Committee for Biology and Medicine, Luigi Califano, stood out for his authority and prestige. A few years later, the reins of the CNR were taken over by Vincenzo Caglioti. Professors Califano and Caglioti both showed enthusiasm at my initiative. It was Caglioti who in 1969 decided to transform the small Center of Neurobiology into an official organ of the CNR, which became known as the Laboratory of Cell Biology. In addition to neurobiology, it included three other departments: cell biology, mechanisms of gene expression, and immunology.

Our research at the Istituto Superiore di Sanitá proceeded with great impetus and success. In the pervasively cheerful and optimistic atmosphere of the first half of the 1960s, reflected in Rome's limpid and luminous air, each new result elicited greater hopes, and the project of deciphering the NGF's mechanism of action seemed near to completion. Even though later, our optimism turned out to be unfounded, it did spur us to overcome the difficulties we were facing as a result of the scarcity of our means and to seek novel approaches. The problem itself seemed to encourage this latter tactic, so vast were—and still are—the unexplored areas of the NGF target cells and mechanism of action.

During the 1960–70 decade, the field of NGF studies was a sort of private hunting ground. Despite the interest aroused by the two articles Stan Cohen and I published in the *Proceedings of the National Academy of Sciences* in 1960, few scientists ventured down our same path, the results being so perplexing and hard to reconcile with prevailing theory.

The creation of the Laboratory of Cell Biology, with whose direction I was entrusted, heralded, however, the onset of new difficulties. The laboratory had to be temporarily set up in a building only a few hundred yards from the Piazza del Popolo, in the very middle of Rome's chaotic traffic. From the very start I felt some discomfort there, accustomed as I had become to the quiet environment of Washington University's campus and to the sense of security and stability engendered by it and other great American universities, not least because of security guards, whose presence permits one to engage in scientific activity at any hour of the day or night. In Rome we were constantly under pressure from the owners of the building, who wanted us to leave to be replaced with higher-paying tenants, and from Public Health authorities who regularly threatened to bar us access to the labs because of possible accidents (which never, in fact, occurred) resulting from the use of radioactive materials in a densely populated area of the city. The situation gradually worsened because, from the second half of the 1970s to the beginning of the 1980s, the CNR demonstrated a total lack of interest in neurobiology which, at the time, was considered a secondary sort of discipline compared with others, molecular biology in particular.

Our group's frustration was increased by the fact that, precisely in those years, NGF studies abroad had soared, while our production

196

was quantitatively if not qualitatively far inferior to that of other European and American laboratories, where an ever-increasing number of scientists worked together in large teams with perfectly equipped laboratories at their disposal. What is more, the group of researchers and technicians that I had put together since before the founding of the Laboratory of Cell Biology was now beginning to lose its members. The notoriety NGF enjoyed was for many of them sufficient reason to delve into other problems. The desire to work independently in the scientific realm—a desire much stronger in Italy than in the United States—causes a withering of productive forces, and scatters them into a thousand separate rivulets. The autonomy that researchers and technicians employed by the CNR enjoy, with respect to the department heads or general director, further kindles the tendency of individuals not to work in teams. The activities of individual researchers or technicians are never the object of control; and, on the basis of the dubious egalitarian criteria adopted by state or para-state institutions such as the CNR, quality of production has no bearing on career advancement or salary levels. The only positive aspect of such a form of regulation is that in Italy scientists are not under pressure to publish, as their American colleagues are, in order to obtain the continuation or renewal of the grants on which they depend. Italian scientists can therefore allow themselves the luxury of tackling "risky" problems which, in the majority of cases, are destined to lead nowhere but once in a while succeed in charting a new path, allowing an individual's creativity to express itself fully.

Often, in the course of those years, I asked myself whether it was the time to desist. If our work remained of a much higher standard than working conditions allowed, it was all the merit of the members of the group, who manifested the dedication and ability to overcome all obstacles. Vincenzo Bocchini, an able young Neopolitan chemist, was one who worked with the group from 1963 to 1969. With imperturbable calm he confronted the increasing difficulties we were having for want of funds and perfected techniques for purifying NGF from mouse salivary glands and identifying it in a molecule that was free from contaminants. The method he developed along with Pietro Angeletti enabled Ruth Hogue Angeletti and Ralph Bradshaw from Washington University to elucidate the amino acid sequence of the protein molecule in 1971. Twelve years later, this knowledge of the NGF's primary structure led to the identification, on the part of two teams of investigators in the United States, of the DNA that codes for the molecule and of its originating gene in different animal species including man. In spite of these important contributions, it would have been very difficult for NGF research to continue in Italy had it not been for two young friends, my collaborators for over twenty years in St. Louis and in Rome, Pietro Calissano and Luigi Aloe. A graduate of Genoa University Medical School and an excellent biochemist, Calissano from the very start set about investigating the most difficult problem posed by the Nerve Growth Factor: the question, that is, of its mechanism of action. This problem, which remains the most arduous of all and has yet to be solved as far as many hormones are concerned, is in the case of NGF, further complicated by two factors: the structural characteristics, and the perennial, nonproliferating nature of receptive cells. Nerve cells differ from other types of cells in their axial prolongation and in the complex synaptic apparatus at their fiber endings. The endocellular processes that take place as a result of NGF activity are difficult to study because cells cultivated in vitro and treated with the molecule are necessarily in their final differentiative stage and would simply die in its absence. It is therefore impossible to conduct parallel analyses of the cells in the absence of the factor. The use of a line of murine neoplastic cells known as PC 12, however, has enabled the two American investigators who introduced it, L. Greene and A. Tischler, as well as Calissano with his Roman team and others, to analyze the molecular processes resulting from the addition of NGF to the culture medium. In the presence of the factor, these cells acquire all the characteristics of sympathetic nerve cells, and the factor's elimination does not cause them to die but, rather, to reacquire the morphological and functional characteristics of their original, glandular strain.

The investigations Calissano carried out in the prohibitive conditions reigning in our laboratory during the last two decades have brought an extraordinarily important contribution to the elucidation of NGF's mechanism of action at the molecular level. The friendship and community of scientific interests that characterized our relationship since the start have enabled us both to overcome the moments of most intense discouragement, which for him, at the start of a career, were even more taxing than they were for me, nearing the end of mine.

At the same time, using the type of approach that Claude Bernard,

198

the great French physiologist, used and that I have, since the start, adopted as my own—an approach that upholds the priority that biological problems have over technical ones, which are always changing in accordance with the requirements of the specific questions being posed—Luigi Aloe and I have also charted new territories. His extraordinary dedication, technical aptitudes, and scientific intuition (qualities picturesquely summed up in the expression "a green thumb") were the elements fundamental to the successes that our own exiguous team, composed of only the two of us, has reaped in the last twenty years.

In July 1979, having reached retirement age, I left my post as director of the Laboratory of Cell Biology; since then I have been allowed, if not without opposition, to continue to work in the capacity of a guest in the institute which I had seen born under better auspices ten years earlier.

Even though the ponderous bureaucracy that oversees public sector institutions, fearful as it is of clashing with the unions and constricted by a real want of funds, has for the last two decades stultified scientific activity by not allowing new personnel to be hired, an "Italian style" remedy has been at work to prevent the complete paralysis of research activities in laboratories such as our own. Young graduates and postgraduates affiliated with and/or stipended by the science departments of the universities have been allowed to work in the laboratories and research institutes of the CNR. Pietro Calissano's team is largely made up of such university dependents and includes investigators who are as competent as they are passionate about their work: Antonino, Silvia, Anna-Maria, Roberta, and Bruna, to mention only those whom I run into most frequently every day and who hearten me with their smiling young faces.

It was in an atmosphere of silent and tenacious pursuit of the research begun thirty years before that NGF took everybody once again by surprise with a sudden and unforeseeable gesture. This time, however, it was not only a small group of people at work in Washington University's Rebstock Hall, at the Institute of Biophysics in Rio, at the Istituto Superiore di Sanitá or at the Laboratory of Cell Biology in Rome who were there watching as the factor opened new, unexpected routes. An increasing number of new admirers better equipped than we are for following its tracks are in hot pursuit, predicting the NGF's movements even before they take place. It was thus that from mere "super" on the great stage of the Biological Sciences, NGF suddenly found itself wearing the protagonist's shoes. How to explain such an outburst of activity after the long period of silence which followed its first appearance on stage? If we rule out the possibility that this "miracle molecule" may have suddenly begun to do things that it had not done before, the only plausible explanation is that the always more advanced techniques developed in recent decades and, especially, in our own, applied to an always more attentive scrutiny of the NGF's spectrum of action, have been able to reveal effects which previously had gone unobserved. It has recently been shown, using a variety of techniques and approaches including recombinant DNA technology and genetic engineering, that the cholinergic nerve cells, which play a role of primary importance in the nervous system, are highly receptive to NGF in a manner not substantially different from that of the sympathetic and sensory cells which had first alerted us to its existence. These findings, in turn, have given rise to the hypothesis that an irregular or insufficient production of the molecule, as well as the destruction it may suffer as a result of any number of physical disorders, may be at the root of many of the nervous dysfunctions for which there is still no known cure. Even if this is no more than a hope and a possibility, it is one which has given enormous impulse to research aimed at either confirming or refuting it. A first, very important result has been the identification of the gene that codes for murine NGF, a finding which was simultaneously reached by eminent experts in genetic engineering in two American laboratories. The subsequent identification of the gene coding for human NGF has, in turn, made it possible to synthesize the latter in great quantities, just as is done with other molecules with important biological functions, such as certain hormones and interferon, for therapeutic purposes.

Will the NGF derived from this new source—and now no longer collected from neoplastic tissues (whose reasons for producing it, by the way, are still cloaked in mystery) or from the mouths of snakes and mice, but, aseptically distilled in the laboratory—be able to bring back order to the functionally impaired neuronal circuitries of that immensely complex entity, the brain of *Homo sapiens*?

A recent communiqué published by a famous Swedish institute states: "The discovery of Nerve Growth Factor (NGF) in the beginning of the 1950s is a fascinating example of how a skilled observer can create concept out of apparent chaos." The chaos alluded to derived from the vagueness and confusion of the notions, in the first half of the century, concerning the mechanisms of differentiation and the functioning of neuronal cells and circuits.

It was in the anticipatory, pre-Carnival atmosphere of Rio de Janeiro that in 1952 NGF lifted its mask to reveal its miraculous ability to cause the growth, in the space of a few hours, of dense auras of nervous fibers. Thus began its saga.

On Christmas Eve 1986, NGF appeared in public under large flood-lights, amid the splendor of a vast hall adorned for celebration, in the presence of the royals of Sweden, of princes, of ladies in rich and gala dresses, and gentlemen in tuxedos. Wrapped in a black mantle, he bowed before the king and, for a moment, lowered the veil covering his face. We recognized each other in a matter of seconds when I saw him looking for me among the applauding crowd. He then replaced his veil and disappeared as suddenly as he had appeared. Has he gone back to an errant life in the forests inhabited by the spirits who drift at night along the frozen lakes of the North, where I spent so many solitary, enchanted hours of my youth? Will we see each other again? Or was that instant the fulfillment of my desire of many years to meet him, and I have henceforth lost trace of him forever?

CHAPTER 23

Farewell to a Master— and to a Father

OUR LAST PARTING took place one late January evening in 1965, at Turin's San Giovanni Hospital, where Giuseppe Levi had been a patient for a few days and, that morning, had undergone radiological examinations revealing the presence of a malignant tumor protruding into the gastric cavity. I entered the small room, whose feeble electric lightbulb was not even lit, and saw the outline of his large body lying cramped on the small hospital bed in the twilight of the winter sunset. He was alone. Glad of my unannounced visit, he greeted me with the gruff voice that, sounding the length of the dark corridors in the Anatomical Institute, had often caused me to tremble. "I have a carcinoma of the stomach," he said, adding its precise location: the pylorus. "I diagnosed it myself," he continued, happy with this evidence of his clinical acumen. "I'll die in two weeks. I don't regret it. On the contrary, I'm happy for it. I've already lived too long." He had turned ninety-two a few months before. His voice didn't betray the slightest emotion, but a painful contraction in the face revealed that he was suffering. Without attempting to question the diagnosis and his certainty that the end was so near—thirty-three years of experience had taught me how useless it would have been to do so—I asked him if he was in pain. "Yes," he answered. "Not so much because of my stomach as from the gangrene in my right foot and the bottom third of the leg." His left leg had been amputated above the knee for the same reason—circulatory deficiency—seven years previously, when he was eighty-five.

He, who had always had a passion for climbing and walking in the mountains, had accepted the grave impairment with a stoicism that

Farewell to a Master—and to a Father

filled us with admiration. About a year before, he had lost his adored life-companion Lidia, the victim of a heart attack, and the psychic suffering caused him by the sudden loss of the person most dear to him had been mitigated only by the onset of pain of a physical nature. Showing the same tenacity he put into all his endeavors, he learned to walk again with the help of an artificial leg and a stick, and went up and down stairs, disdainfully refusing the aid offered by assistants and friends, whom he would push aside whenever they tried to offer help. Neither old age nor the amputation of a limb dissuaded him from continuing every summer to go to a small village in the Swiss Alps which he loved for Lidia's sake, or from making the train journey every few months from Turin to Rome to attend meetings of the Academy of the Lincei where, thanks to his authority, his opinion continued to prevail regarding the directions to be taken in the academy's scientific activities. He traveled alone. To friends and assistants who worriedly asked how he managed to climb down from trains and the tiring haul along the station platforms with his suitcases, he answered impatiently, "Like everybody else. When I'm tired, I ask one of the porters to carry me on one of his baggage carts. They're somewhat hard but very comfortable."

I sat down next to his bed. "Tell me about your work," he demanded, breaking the silence which was weighing upon us. I obeyed. I told him that I was going back to Rome the next day and would leave the day after for St. Louis, where I would immediately resume my teaching and research. My work was in a particularly promising phase, because of all the confirmation of our findings coming from other laboratories, and because of the possibilities I saw for new developments. He wanted to know, down to the slightest detail, about the results I had obtained in recent months, and revealed a perfect understanding of those results whose importance he had at first underestimated. To my relief, no one came to disturb that last conversation of ours, which went on for three hours in the dark while he held my hand tight in his own, still robust and strong as that of a young man. We reminisced about what we had enjoyed and suffered together in the past, about my anguish at not having been able to unravel the mechanisms behind the cerebral convolutions in the brain of human fetuses. He smiled, recalling those times, and squeezed my hand to show he had forgiven me my youthful ineptitude. After those years, there had followed the racial persecutions,

203

our dismissal from our respective posts, the Belgian interlude, the partial resumption of scientific activity in my small home laboratory, and, finally, the time spent in Florence where, by some lucky coincidence, we had both sought refuge during the Nazi occupation. We spoke of his pupils and my classmates; of my cousin Eugenia, who despite the difficulty of being a wife and a mother had courageously pursued her research career in Buenos Aires, where she had gone to stay with her husband and children at the start of the anti-Semitic campaign. We spoke of Salvador Luria, Renato Dulbecco, and Giovanni Godina, one of Levi's disciples who became professor of veterinary sciences at the University of Turin, and to whom Levi was bound by a sense of esteem and deep affection after the many years the former had spent keeping him up to date on the results of in-vitro experiments he had begun with the Master; of Guido Filogamo, the youngest among us, who was involved as I in neuroembryological research and Levi's direct successor in the chair of anatomy. Finally we spoke of his favorite pupil, Rodolfo Amprino.

Night had fallen. The room was still dark, illuminated only by the light of the streetlamps outside the window that gave onto the garden in front of the hospital. I asked him if he had requested a night nurse. "No," he answered, "I don't need any help." With some difficulty, I persuaded him to allow me to try to get hold of a nurse whom I knew and respected. He grumbled out some form of consent and, till the end, had the attention he reckoned superfluous. He kissed me when he dismissed me. I had a lump in my throat and could neither talk nor abandon him. He cut things short, saying without melancholy, "This is our last farewell. I thank you, Rita, for what you have done and hope your good fortune continues."

Two weeks later, in St. Louis, I received news of the death he had foreseen with such detachment and lucidity. Twenty-three years after that conversation—which differed from previous ones only in the pain of the shared certainty that it was to be our last—and more than half a century since I made my first entrance into the anatomical amphitheater where his figure towered over the corpse, I relive experiences marked by my daily contact with a man of his extraordinary temperament and personality. Despite the fact that he differed in many respects from my father, certain traits they had in common had struck me from the start. They had the same resolute and imperious way of dealing with people and problems, the same energy which

Farewell to a Master—and to a Father

knew no obstacles to the realization of their work projects, and, finally, the same intolerance for superficiality and for the failure to perform duties. Both would burst into terrible fits of rage, shattering for the guilty parties, which would be quelled as quickly as they had begun. But unlike my father's life which had been tumultuous and painful, Levi's had been calm and ordered, and entirely taken up by scientific activity, university courses, and, what he preferred above all, participation in the research projects he assigned to his pupils. The rhythm of his daily work, which started at seven in the morning and ended at eight in the evening, was interrupted only by the long vacations he allowed himself, to engage in his favorite pastime: excursions in the mountains with Gino, his son who resembled him extremely.

Our farewell—in that small hospital room whose bareness reminded me of a Franciscan's cell—could not have taken place in surroundings or under conditions more apt to bring out his qualities as a man and a scientist, consumed neither by old age, nor suffering, nor by the knowledge he had of his approaching end. He accepted the latter with a stoic serenity while maintaining till the very last an interest in research as an instrument for the understanding of nature and not as an object of competition and an instrument of power. In an era when the latter conception of scientific activity prevails, that last conversation of ours, the vivid interest he took in what he knew must be his last "briefing" on the state of my research, revealed to me the secret of the great influence he exerted on the young. It derived from the passion with which he had pursued his studies and later directed those of his pupils, while remaining indifferent to the honors and plaudits that are granted old Masters.

At the age of ninety-two, Giuseppe Levi was still too young and full of interests to be pleased by such fatuities—as, with supreme disdain, he was wont to call them.

CHAPTER 24

Disharmony in
a Complex System

As a product of evolution, man is only rough-hewn; he lacks the biological polish that comes from a long slow adaptive improvement through natural selection.

—THEODOR DOBSHANSKI

IF a passionate dedication to the study of the development of the nervous system and ensuing investigations of the specific growth factor of certain lines of nerve cells had absorbed my life's most productive years, another aspect of the same system—namely, its function—was to become in more recent times the object of my intense interest and meditation. During my first years at the university, I had often contemplated the brain—preserved in formaldehyde on a dusty shelf of the museum of the Anatomical Institute of the University of Turin—of a renowned anatomist who had requested that upon his death his brain be preserved and used as an object of study in the same institute where he had taught for half a century. Given the cynicism and indifference of the young, to whom death appears to be an event so remote as to not warrant consideration, the brain had become an object of joking commentary rather than of respect. The anatomist, a faithful follower of the criminologist Cesare Lombroso, had in his lectures described a circumvolution he believed was especially developed in criminals. The same circumvolution—the students maintained—was plainly visible in his brain, and they would point it out to each other under the disapproving eyes of the janitor, who had loved and respected the old professor.

206

Disharmony in a Complex System

In a different spirit and state of mind, the neurologist Eric Harth tells, in his interesting book *Windows on the Mind* (1982), of the reaction of one of Einstein's great admirers, Steven Levy, upon seeing a jar containing the brain of the famous physicist: "I had risen up to look into the jar, but now I was sunk in my chair speechless. My eyes were fixed upon that jar as I tried to comprehend that these pieces of bumps hobbling up and down had caused a revolution in Physics and quite possibly changed the course of civilization."

What mysterious sequence of mutations caused the brain of our distant predecessors so to increase in volume, and its cortical mantle to expand, as if by levitation, and fold back upon itself in ever-more creases, or circumvolutions, as to fit within the limits set by cranial capacity, unvaried over hundreds of thousands of years? While it is possible, on the basis of the inspection of skulls that have been uncovered and dated, to reconstruct the brain's progressive increase in overall volume over eras geological ages apart, from the late Miocene to the Pleistocene, it is not possible to reconstruct the process that increased the number of circumvolutions. This process has left no trace. Refuting teleological concepts which, before the Darwinian revolution, saw in the appearance of the human species the ultimate end of all creation, the principle of the irreversibility of evolutionary processes acknowledges that, once begun, each follows the others along a unidirectional path. This principle has now become universally accepted. Subsequent mutations further increase the advantages acquired as a result of preceding ones. The mutations that caused, in the course of past eras the progressive increase in the volume of the brain, and in the complexity of the neuronal circuitry connecting the billions of nerve cells (whose number, especially in the cortical areas, also increased), were followed by other mutations which endowed our predecessors with cognitive capacities lacking in their contemporaries and in descendants of other vertebrates. In the three and a half million years since the time our ancestress "Lucy," whose cranium was roughly the size of a coconut, met her death in what is today the Afar triangle in the Hadar desert north of Ethiopia, these capacities became manifest in her descendants, whose greater height and cranial volume are apparent from findings made in geologically more recent quarries. According to one of those mysterious series of mutations which mark the evolution of all living species, that to which the small biped Lucy belonged dissociated itself from the other hominoid species, with whom it had shared millions of years of tree life in the African jungles, and ventured into the perilous open spaces of the savannah. Of Lucy's direct or indirect descendants— belonging to the genus *Homo habilis,* who lived in what today are Ethiopia and Tanzania between three million and one million seven hundred thousand years ago—paleoanthropologists have found vestiges, such as splinters of stones cut about three million years ago and of stones transformed into rudimentary defensive and offensive weapons in more recent epochs, which reveal the progressive refining of an ability to produce artifacts. The uncovering of the bones and skulls of individuals belonging to *Homo habilis,* in scattered quarries over one million years old throughout the African continent, already attested to an inclination that brought the species to migrate beyond the confines of Africa into Eurasia.

It is impossible to determine when our species' unique ability to communicate by means of that extraordinary tool that is language actually developed. The fundamental difference between language and the means of communication of other animal species lies in the former's cognitive and non-emotive content—a result of the development of the circumvolutions governing vocalization and the elaboration of information received through the afferent visual, acoustic, and other sensorial systems. Possession of such faculties undoubtedly played an essential role in the survival of individuals who, thanks to language, were able to confront stronger and more agile predators in alliance with other members of their own species, to be informed of the presence of the enemy along their nomadic routes. It is still not known when *Homo habilis* or his descendant *Homo sapiens Arcaicus* developed either this vocal capacity or the much more elaborate one of forming thoughts. The latter has been documented by the findings, in geologically more recent quarries, of ritual objects, which point to the existence of cults of the dead and are the unequivocal index of an individual and tribal self-awareness. From nomadic life, these forebears of ours—or, at least, a large percentage of them— turned to sedentary life, a necessary condition of the development of primitive forms of social organization founded on norms transmitted orally from parents to children. This last great leap forward in the evolution of our species is believed to have occurred about forty-five thousand years ago. Since then, or since about one hundred thousand years ago as some anthropologists maintain, increases in intellectual capacity have ceased to be the result of either chance mutations or simple natural selection, and the survival of Lucy's recent descendants has no longer been entrusted to the slow adaptive processes of biological evolution. Rather, it came to be guaranteed only by the advent of cultural evolution, which gave vent to new and more efficient strategies for the consolidation and propagation of the *Homo sapiens* species. Essential among the latter was the discovery that knowledge could be transmitted not only by means of a spoken language, but also by way of writing—a discovery it is believed was made about ten thousand to twelve thousand years ago. Since then, possession of the cultural patrimony accumulated by distant and recent ancestors has entailed an enormous increase in the rate of absorption of new acquisitions and radically transformed the living conditions of succeeding generations, to herald a new era in human history.

The biological evolutionary processes that brought new circumvolutions into being and set the bases for cultural evolution did not, however, affect in equal measure the structural and functional developments of the archaic components of cerebral circumvolutions, nor did they much alter that constellation of interconnected subcortical nuclei known as the limbic system, which governs the processing and expression of emotive and affective activities. The evolutionary disparity between the cortical areas of recent formation, where cognitive functions are processed, and those of the limbic system may be explained by the fact that the latter have never ceased to carry out, in both lower and upper mammals (man included), the same tasks, vitally important in respect to the perpetuation of the species and the role of individual offensive and defensive mechanisms. By virtue of its fundamental properties, the limbic system did not lend itself to the random game of mutation, nor has it in any way been influenced by cultural evolution, which affects only circumvolutions of recent formation.

The portentous increase in cognitive capacities following the discovery of symbolic systems of written and oral communication is amply evidenced by the architectural, artistic, and literary works that have reached us from East and West alike, all of which are accomplishments of recent millennia. But it was only the discovery of differential calculus three centuries ago, and of its applications, that engendered the rapid technological development that, in our time, has enabled human beings to land on the desolate plains of the Sea of Tranquility on the moon, and led to the fissioning of atoms and our realization of the formidable energy born out of such a process. The vertiginous growth of the constructive and destructive powers of *Homo sapiens-sapiens*—in striking contrast with the invariability of human emotive faculties, which, today as in the past, control conduct and actions—is the prime cause of the dangers that threaten the very survival of our species. In fact, if since the dawn of civilization neither the ethical principles embodied in different religions nor rational capacity have been able to prevent wars, massacres, and genocide, today the possession of forces of a destructiveness without precedent and our inability to control them are dangers threatening man, as we are all too aware.

Is the perpetuation of war and of slaughter—fatal activities unique to our species—the ineluctable consequence of an irreducible, genetically transmitted aggressiveness, as is maintained by etiologists and sociobiologists, who in the last couple of decades have enjoyed a broad consensus? Or is it the result instead not only of biological but also of social and cultural factors which in our species alone condition the conduct of individuals and of the masses? This latter hypothesis is upheld by competent neurologists and biologists who are also connoisseurs of human nature: for example, Salvador Luria, who expounds it in his excellent book *Life, the Unfinished Experiment,* and among dozens of others, W. H. Thorpe, and Ashley Montagu. The problem was analyzed in all its multiplicitous aspects by the well-known writer Arthur Koestler, who took into account both sociocultural factors and the disparity between the cerebral systems governing cognitive and emotive faculties. The young of our species differ from those of other mammals in their slow physical and intellectual development, a slowness that makes them depend on their parents or guardians during the long period between birth and puberty. Though the splendor and complexity of the brain of *Homo sapiens* is due, in large part, to the slowness of the process of maturation of intellectual faculties, a protracted dependency on adults also leaves indelible marks on the nervous structures governing individual behavior. Koestler states that the imprinting period, which in ducklings and other mammals lasts only for the first few days or weeks of life, in humans extends not only until puberty but for the whole of life's duration, from the cradle to the grave. The ethico-social systems that individuals are exposed to while young, whether they belong to tribes isolated from the rest of civilization or to more evolved contemporary societies, condition their adult behavior as well, engendering tight bonds between members of the same ethnic group, who are united by a credo and ready to sacrifice everything in order to defend blindly accepted values. The modern day soldier, like the Roman legionary of old and the host of unknown soldiers who over the centuries have been caught in war's fatal grip, owes his condition as potential killer and victim not to any innate ferocity or aggressive instinct but, rather, to his passive acceptance of orders.

Yet, if esprit de corps and blind obedience cause millions of individuals to set themselves at the disposal of their governors for the purpose of waging war, other more subtle and harmful forces come into play to transform them into enthusiastic followers of tyrants, ready to slaughter millions of other individuals who, just like them, may be drunk on slogans and hypnotized by the power that symbols can exert on the human brain. To those who have still before the mind's eye the image of the interminable ranks of the Hitler Youth, transfigured as they were by their love for the Führer and prepared to sacrifice everything to the whim of their deity, the magic of symbol, word, and emblem becomes manifest in all of its terrible strength.

In this sinister way, language, the greatest privilege bestowed on human beings, casts them into an abyss of obscurantism whenever they fall prey to fanatical and cynical leaders who use words to incite hatred, or when a swastika or the hood of a Ku Klux Klan member, evokes mystic reactions that inhibit the intellectual faculties of that distant descendant of the small biped who found her way to us and was baptized, three and a half million years after her birth, with the name Lucy.

Epilogue:
Primo Levi's Message

I beg the reader not to go looking for messages. It is a term which I abhor because it puts me in crisis, because it dresses me in clothes which are not mine, which on the contrary belong to a human type whom I view with distrust: the prophet, the bard, the soothsayer. That I am not.

—Primo Levi

IF by the term *message,* one means the contents of an oration propounded from a pulpit, or a harangue directed from a balcony to a crowd packed in a public square, or a passionate appeal to the emotional and irrational impulses that since the dawn of civilization have guided the behavior of man in times of crisis and danger, then you, Primo, certainly do not belong to the group of those who have assumed unto themselves the right to deliver such messages. But the mystical-theological definition of the term, which you refer to and refuse to be associated with, dates back to a time when both natural phenomena of which the causes were unknown and the deviant or normal behavior of individuals were attributed to the inscrutable will of the gods, a will revealed by prophets, bards, and soothsayers. Today the terms *message* and *messengers* have lost their sacred aura and have acquired a human and secular dimension. With the extension of their meaning, they have come to be used also to indicate chemical agents that transmit information from one cell to another. These "messages," devoid of either divine or anthropomorphic connotation, play a fundamental role in organic processes and especially in those thought to be typical of the nervous system.

Epilogue: Primo Levi's Message

Your message, Primo, which tens of thousands of readers have received, belongs neither to the first nor to the latter of these categories. It shares with those of the first no tribal or barbarian connotations, and partakes of none of the emotional detachment and scientific rigor of the second. Its extraordinary force cannot be attributed either to a prophetic tone—which, in all instances, you disdainfully avoid—or to the novelty of its content. Others before you have, and others after you will continue, to denounce the tragic consequences of servile devotion, unconditioned obedience, and supine acceptance of orders from fanatical and paranoid leaders. But no one has ever done it with the same suffered efficacy as you, with a more implacable analysis of the mentality and motives that led criminals, such as the commandant at Auschwitz, to act as they did; or, at the same time, with a greater detachment and absence of hatred. And no one has ever denounced as you did "with what great ease good can yield to evil, be besieged and finally overcome by it"; or how it can, in the case of the criminals whom you scrutinized, "survive in small grotesque islands, ordered family life, love for nature, a Victorian morality." Rudolf Höss, the commandant at Auschwitz—"one of the greatest criminals of all time," as you define him—"was not, however, made of a different stuff from that of any other bourgeois of any other country; his guilt, which was not written in his genetic patrimony or in the fact of having been born a German, consisted entirely in his not having known how to resist the pressure that a violent environment had exerted upon him even before Hitler's rise to power."

Is it justifiable to weave, out of an awaremess of these aberrations in the behavior of *Homo sapiens-sapiens,* an elegy to imperfection? Is it justifiable to celebrate the dominant characteristic of human activity: both during critical periods in history when—as you have described—the lives of thousands have been at the mercy of obsequious and cruel servants of a despot; as well as during those other periods when life proceeds with relative tranquillity, and only momentarily does vanity, hunger for power, or mere abjectness in the face of authority make visible the less valuable side of human nature? And if the evolutionary disparity between the intellectual and the emotional faculties of our cerebral centers is the result of biological processes for which we are not responsible, should we for this reason feel heartened to belong to a species well endowed with higher cognitive abilities yet, in its conduct, tragically governed by the dominant

213

emotional ones? However, for the millions of individuals who bear the stigmata of this disharmony, who belong to what you call the "gray zone" and who out of cowardice collaborate with their executioners, there have been thousands of others throughout the ages who have not surrendered—either in the face of torture or of death—and who have kept alive the fire of hope for their comrades as well. You were one of them, Primo. You who explained to your young comrade Jean—who couldn't understand Italian and had been cast along with you into the inferno at Auschwitz—the meaning of Ulysses's admonition, which sounded for you like the very voice of God:

> Considerate la vostra semenza:
> Fatti non foste a viver come bruti, . . .*

Your exhortation to Jean, to all of us, and to those who are yet to be born, has miraculously overcome the thousand obstacles opposing its diffusion and your survival. Many have received it and many more, it is my hope, will reap its wisdom in the future. Yours is a secular "Spinozan" message, full of a sad, deep awareness of the harm one human being can inflict on another. But, at the same time, it is a message of hope, because whoever has voiced it while in the deepest despair, as were you, has kept intact the highest qualities of *Homo sapiens-sapiens* and come out of the most atrocious of all experiences with an upright forehead and a spirit pure.

---

*"Take thought of the seed from which you spring:
You were not born to live as brutes, . . ."

NOTES

PAGE

11–12   Giorgio De Chirico, *Paola Levi Montalcini,* (Turin: Accame, 1939).

26   François Poulain de La Barre, *De L'Egalité des Deux Sexes Corpus des Oevres de Philosophie en Langue Française,* (Paris: Librairie Arthème Fayard, 1984 [1673]) "Preface Originale," p. 10.

27   John Ruskin, "Sesame and Lilies," in *Sesame and Lilies: The Two Paths and The King of the Golden River,* (New York: E. P. Dutton, 1960), pp. 59, 60, 63, 65.

27–28   John Stuart Mill, *The Subjugation of Women,* (Cambridge, MA: M.I.T. Press, 1970 [1869]), p. 76.

47   Morris West, *The Devil's Advocate,* (London: Heinemann, 1961), p. 1.

59   Alexis Carrel, *Man, the Unknown,* rev. ed., (New York: Harper & Row, 1939).

74–75   Matteo Matteotti, *Quei Vent'Anni,* (Milan: Rusconi Editore, 1985), pp. 11–26.

77   Giacomo Leopardi, "Canzone all'Italia," verse 4.

80   Renzo De Felice, *Storia degli Ebrei Italiani Sotto il Fascismo,* (Turin: Einaudi Editore, 1972), p. 255.

80–81   Ibid., p. 450.

81   Etty Hillesum, *Diario: 1941–43,* (Milan: Adelphi Editore, 1985), p. 167.

81   De Felice, *Storia degli Ebrei,* p. 155; (Ovazza citation), p. 424.

84   Rita Levi-Montalcini and Fabio Visintini, "Relationship Between the Functional and Structural Differentiation of the Nerve Centers and Pathways in the Chick Embryo," *Archives Suisses de Neurologie et de Psychiatrie,* Zurich, 1939, pp. 1–45.

88   Enzo Biagi and Sergio Zavoli, *Dieci Anni della Nostra Vita,* "Mussolini's War Proclamation of 10 June 1940," sound recording.

89   Dante, *Inferno,* Canto xxvi, lines 121–123.

91–92   Benvenuta Treves, *Tre Vite,* (Turin: Case Editrice Israel, 1954), p. 174.

97   Roberto Battaglia, *Storia della Resistenza Italiana,* (Turin: Einaudi Editore, 1983), p. 81.

Notes

100   Ibid., p. 86.

123   George Bishop, *Annual Review of Physiology* 27 (1965): 3.

123   Francis Crick, *Of Molecules and Men,* (Seattle: University of Washington Press, 1967), p. 74.

124   Ibid., p. 99.

124   Ibid., p. 22.

125   Gunther S. Stent, "The Nervous System," in *Paradoxes of Progress,* (Oxford, Eng.: W. H. Freeman, 1978).

127   Camillo Golgi, "La Dottrina del Neuron: Teoria e Fatti," Nobel Lecture, 11 December 1906..

128   S. Ramón y Cajal, *Recollections of My Life,* trans. E. Horne-Craigie, (Philadelphia: American Philosophical Society, 1937), pp. 323–25.

128–29   M. Abercrombie, "Ross Granville Harrison," *Memoirs of Fellows of the Royal Society* 7: 113.

158   Alexander Luria, "Neurophysiology: Its Sources, Principles, and Prospects," in *The Neurosciences: Paths of Discovery,* ed. F. G. Worden, J. P. Swazey, and G. Adelman, (Cambridge, MA: M.I.T. Press, 1975), p. 339.

178   Dylan Thomas, "Lament," in *In Country Sleep and Other Poems,* (New York: New Directions, 1952).

192   Robert Walgate, "Can Order Spring from Chaos? Science in Italy," *Nature* 30 (1983): 109–18.

207   Erich Harth, *Windows on the Mind: Reflections on the Physical Basis of Consciousness,* (New York: William Morrow, 1982), p. 13.

210–11   Arthur Koestler, "The Urge to Self-Destruction," in *The Place of Value in a World of Facts,* Proceedings of the 14th Nobel Symposium, Stockholm, pp. 297–303.

213   Primo Levi, "Preface," Rudolf Höss, *Comandante ad Auschwitz,* (Turin: Einaudi Editore, 1985).

INDEX

Abercrombie, M., 128–29
Adrian, Edgar, 132
Aloe, Luigi, 198, 199
Ambrosio, Vittorio, 100
Amprimo, Rodolfo, 51, 54, 65, 89, 111, 113–14, 117, 204
Angeletti, Piero, 168, 179, 194, 197
Angeletti, Ruth Hogue, 197
Anna (aunt), 23, 24, 25, 36, 37, 47
Antonietta. *See* "Cincirla"
Artom, Emanuele, 91, 92
Ascoli, Guido, 39

Badoglio, Pietro, 97, 100
il Barba (great uncle), 21–22, 25
Barsotti, Renato. *See* "Neroncino"
Bernard, Claude, 198–99
Bishop, George, 123, 133, 170–71, 175, 180
Bishop, Michael, 158
Bocchini, Vincenzo, 197
Boeke, Jan, 127–28
Bottai, Giuseppe, 56
Bradshaw, Ralph, 197
Broca, Paul, 129
Brontë, Emily, 15, 35, 36
Bruatto, Bettina, 36
Bruatto, Caterina, 36
Bruatto, Giovanna, 36–38, *photo*
Bueker, Elmer, 144–45, 146
Buzzati-Traverso, Adriano, 193

Caglioti, Vincenzo, 195
Califano, Luigi, 195

Calissano, Pietro, 198, 199
Carlo (chauffeur), 37
Carlo Alberto, King, of Piedmont and Sardinia, 13
Casorati, Felice, 35, 187
Celeste (cousin), 29
Chagas, Carlos, 152, 153–54
Chouteau, August, 135
Ciano, Galeazzo, 69, 76, 103
"Cincirla," 19–20
Cohen, Stanley, 161–68, 196
Colonna, Vittoria, 32
Conti (medical school janitor), 49, 58
Costanza (aunt), 31
Crick, Francis, 123, 124, 125, 193
Cuénot, N., 6, 7

D'Annunzio, Gabriele, 57
De Bono, Emilio, 79
De Chirico, Giorgio, 11, 12, 111, 186
De Felice, Renzo, 80, 81
de Lagrange, Léon, 6
Delvaux, Paul, 119
De Stefani, Alberto, 56
De Vecchi di Val Cismon, Cesare Maria, 56
Diaz, Armando, 34
di Rienzo, Cola, 109
Dobshanski, Theodor, 206
Dulbecco, Renato, 49–50, 52, 65, 112–13, 137, 140, 151, 204
Dumini, Amerigo, 74–75

Eccles, John, 133
Einstein, Albert, 7, 31, 91, 207

Index

Emanuele (uncle), 23
Emanuele Filiberto of Savoia, 11
Erlasser, Walter, 124
Eugenia (cousin), 39–40, 49, 52, 54, 55–56, 57, 59, 60, 204

Farinacci, Roberto, 79
Fazio, Cornelio, 50, 52, 65, 66
Federzoni, Luigi, 56, 75
Fermi, Enrico, 125
Fernando. See "Nando"
Firquet, Professor, 85
Fischer, Emil, 153
Formiggini, Angelo, 82
Freud, Sigmund, 4, 181
Friedmann, Leonore, 194
Fubini, Gina, 82–83, 110
Fuhrmann, Martha, 194

Garibaldi, Giuseppe, 11
Gennaro, San, 148–49
Giorgio (cousin), 32
Godina, Giovanni, 204
Golgi, Camillo, 126, 127
Gray, Elizabeth, 139
Greene, L., 198
Guido (friend), 63–64, 92–93, 108, 172
Gurvich, Alexander, 61

Hamburger, Martha, 120–21
Hamburger, Viktor, 93, 94, 113, 117, 120–22, 130, 131, 138, 139, 140, 144, 145, 146, 152, 154, 155, 156, 158, 161, 162, 166, 167–68, 170–71, 173, 176, 179, 180, 188, 194, photo
Harrison, Ross Granville, 59
Harth, Eric, 207
Henny (friend), 86
Hillesum, Etty, 81, 83
His, Wilhelm, 127

Hitler, Adolf, 7, 25, 80, 87, 91, 93, 104, 109, 153, 211
Höss, Rudolf, 213
Hoyle, Fred, 125

Interlandi, Telesio, 79

"Johnny" (colleague), 3–4

Koestler, Arthur, 210–11
Köhler, Wolfgang, 129
Kornberg, Arthur, 164, 173

Laclède, Pierre, 135
Lagerlöf, Selma, 15, 35
Landau, Bill, 171
"La Pasionaria," 91
Laruelle, Professor, 68, 85
Lashley, Karl, 129–31
Leoncini, Consilia, 102, 103, 104
Leoncini, Cosetta, 102, 104
Leoncini, Ernesto, 102, 104
Leopardi, Giacomo, 76
Levi, Alberto, 106
Levi, Gino, 104
Levi, Giuseppe, 48, 49, 50, 51, 52–7, 58–61, 63, 64, 65–66, 76, 85, 86, 91, 93–94, 95, 104, 111, 112–13, 127, 149–50, 152, 153, 202–5, photo
Levi, Leo, 77
Levi, Lidia, 77, 104, 203
Levi, Mario, 77
Levi, Paola, 104
Levi, Primo, 212–14
Levi-Montalcini, Adamo (father), 4, 13, 14, 15, 16–17, 19, 20, 21–23, 24–25, 28–31, 34, 36, 37, 38, 42–47, 80, 109, 185, 204–5, photo

Index

Levi-Montalcini, Adele (mother), 4, 13, 14, 16, 19, 20, 23, 24, 28–29, 30, 31, 36, 37, 38, 42, 43–45, 46, 47, 67, 90, 99, 102–3, 108, 117, 144, 186, 187–91, *photo*

Levi-Montalcini, Anna. *See* Levi-Montalcini, Nina

Levi-Montalcini, Emanuele, 107, 110

Levi-Montalcini, Gino, 4, 13, 14–15, 16, 28, 32, 35, 43–44, 46, 47, 67, 68, 90, 91, 95, 99, 100, 102–3, 107, 110, 186, 189, 190, *photo*

Levi-Montalcini, Mariuccia, 99, 100, 102–3, 107, 110

Levi-Montalcini, Nina, 14–15, 34, 35, 36, 46, 47, 85, 87, 100, 101, 110, 186, 189, *photo*

Levi-Montalcini, Paola, 11, 14–15, 16, 17, 18, 29, 30, 34–35, 36, 37, 39, 45, 46, 67, 68, 82, 90, 99, 102–3, 105, 117, 144, 185, 186–87, 188, 189, 190, *photo*

Levy, Steven, 207

Li Causi, 66

Lobetti-Bodoni, Professor, 39, 40

Lombroso, Cesare, 206

Louis IX, King, 135

"Lucy," 5, 207–9, 211

Lugaro, Ernesto, 127

Luria, Alexander, 158

Luria, Salvador, 50, 65, 66, 77, 112, 113, 137–38, 140, 142–43, 193, 204, 210

Magri, Gigi, 62

Malpighi, Marcello, 51

Manno (uncle), 28

Marconi, Guglielmo, 136

Marini Bettolo, G. B., 193

Mascagni, Amilcare, 73

Matteotti, Giacomo, 73–75

Matteotti, Isabella, 74

Matteotti, Matteo, 73–76

Matteotti, Velia, 74

Maxwell, James Clark, 192

Mendel, Gregor, 111

Meyer, Hertha, 59, 152, 153, 155

Mill, John Stuart, 27–28

Monod, Jacques, 193

Montagu, Ashley, 210

Moog, Florence, 120

Morgan, Thomas Hunt, 111

Mori, Marisa, 102

Muller, Hermann, 138–39

Mussolini, Benito (the Duce), 53, 56, 57, 69, 74, 75, 77, 80, 82, 87, 88, 91, 97, 103, 109, 111, 136

Mussolini, Edvige, 80–81

"Nando" (friend), 170–81, *photo*

"Neroncino," 73–74

Nitti, Francesco Saverio, 76

O'Leary, James, 171

Oparin, Alexander I., 124

Oscar (partisan), 92

Ottavia (cousin), 29

Ovazza, Ettore, 81–82

Ovazza, Nella, 82

Palmas (medical school janitor), 51

Paolo (cousin), 32

Pende, Nicola, 83

Poulain de La Barre, François, 26

Preziosi, Giovanni, 77, 79

R., Germano (friend), 67–68

R., Luigi (friend), 67

Ramón y Cajal, Santiago, 52, 84, 89, 90, 126–27, 128, 131

Rava, Lucy, 137

Index

Rava, Paolo, 136–37
Rava, Sylvia, 136–37
Reumont, Mlle., 86
Roatta, Mario, 100
Roosevelt, Franklin D., 91
Rooz, Frances, 173
Rosselli, Carlo, 76
Roselli, Nello, 76
Rosetti, Dante Gabriel, 49
Ruskin, John, 27, 31

"S" (friend), 62–63
Sagan, Carl, 125
Salvemini, Gaetano, 76
Sappho, 32
Segre, Sion, 76–77
Selassie, Haile, 91
Severi, Francesco, 56
Sherrington, C. S., 128
Sonneborn, Tracy, 138–39
Spemann, Hans, 93
Sperry, Roger, 130–32
Spiegelman, Sol, 193
Stampa, Gaspara, 32
Stanley, Wendell, 124
Starace, Achille, 79, 82
Stefanelli, Alberto, 132
Stent, Gunther, 125
Stevens, Colonel, 104
Szilard, Leo, 125

Teodoro (uncle), 23, 24
Terni, Tullio, 52–57, 59
Thomas, Dylan, 172, 178
Thorpe, W. H., 210

Tischler, A., 198
Treves, Claudio, 76
Turati, Filippo, 76

Valobra, Nino, 138
Verdi, Giuseppe, 11, 31
Victoria, Queen, 4
Villa, Olga, 102
Villarini, Pasquale, 73
Visco, Sabato, 56, 83
Visintini, Fabio, 84, 90
Vittorio Emanuele II, King, 11, 12–13, 80
Vittorio Emanuele III, King, 53
Volpi, Gioacchino, 56
von Lenhossek, M., 127
von Liebig, Justus, 123

Waldeyer, W., 127
Walgate, Robert, 192
Weiss, Paul, 129, 130–32, 152
Watson, James, 138
Wernicke, Carl, 129
Wholer, Friederich, 123
Wigner, Eugene, 124
Woolf, Virginia, 15

Yeats, William Butler, 5, 169

Zacconi, Ernete, 30–31, 185